Evolution

Evolution

How behaviour prompts elaborative development

Robert E. Culver
JD., PhD.

2019
Brisbane, Queensland, Australia
robert.epigen@gmail.com

Copyright © 2019 Robert E. Culver
All rights reserved.
ISBN:
ISBN-9781795872119

Dedicated to Sabien

*who helped me to understand
that he understood*

CONTENTS

Preface xi

Part One: Evolution 1

Organic continuity 1
Natural kinds 5
Tangible evidence 11
Comparative analogy 16
Expert testimony 20
Positive persistence 37
Constrained pathways 41
Social context 48

Part Two: Life 57

Elusive origin 57
Initial process 66
Heredity 81

Intelligence	86
Organism	90
Insinuation	94

Part Three: Reality — 100

Perspective	100
Perception	106
Meaning	110
Anticipation	115
Communication	117

Part Four: Behaviour — 121

Interface	121
Purposive	123
Appropriate	127
Experiential	129
Obligate	131
Communal	135

Part Five: Recollection — 139

References — 149

Preface

Were it not impudent and therefore imprudent to do so, this essay could be prefaced, summarised, and concluded by the bald statement that we (the scientific community at large) know neither the essence of life nor the processes of its evolution.

Innocence of these details to one side, an evolution occurs when the plastic lifeform decisively surmounts, that is finds fresh congruity with, a persistent or recursive challenge. The defining characteristic of the lifeform is life. The defining characteristic of challenge is an altered reality. Behaviour positions the lifeform in that reality. Part One lightly canvasses particulars raised by the concept of animate evolution. Part Two examines life. Part Three probes reality. Part Four solicits the pivotal role of behaviour in organic change. A final Part Five affords a brief homily to recapitulate and consolidate. To assist the reader navigate in a field of study wherein both the most conventional and the most speculative theories lie equally beyond reprimand of proof, explanatory footnotes are appended as are also extravagant references, the latter not so much to cadge another's authority (but that too) as to signpost the student towards avenues of richer elucidation.

This monograph is part octogenarian vanity and in part scholarly digest, it likely to irritate the conservative but encourage the progressive biologist. The thesis unambiguously rejects the established paradigm of selectionist evolutionary biology, a palpable disenchantment with the schooled 'Atlantic' neo-Darwinism commending, *inter alia*, a contrasting appreciation of the 'Continental' scientific works of Goethe, Bergson, Uexküll, Schindewolf, Weiss and Lima-de-Faria. The essential difference, therefore, between this and most other accounts of evolution is that, rather than concede to a Spenserian notion of the passive organism buffeted by external force, or give credence to a neo-Darwinian adaptionism delineated as the shaping of random genetic error by an unforgiving externality, it is proposed that living organisms have essentially been responsible for their own evolutionary development across time: explicitly, that life is by its effects percontatorial, self-directed by volitional curiosity and adventitious expedience.

Preface xiii

We have inherited a world in which God sits in his Heaven and Darwin sits abreast the main staircase of the British Museum of Natural History. Neither imagery warrants the chore of a piecemeal criticism of either antique explanatory system and little attempt is here made to do so. Instead, a refreshing if unexceptional direction is urged to replace passive notions of evolutionary process by the positive intervention of lifeforms in their own genesis. Much of the literature that already commends this approach unfortunately does so encumbered by the distracting bias of a normative natural selection.

Part One: Evolution

1.1 Organic continuity

The central problem for biology is that we are presently excluded from any comprehensive understanding of lifeforms because we do not know what life is.[1] In the face of such a conspicuous Menoan paradox,[2] to then proceed to speculate why, how, or indeed if, this same unexplained life subsequently evolved into its present multitudinous forms displays an impatience that only the impelling mystery of the subject matter can itself excuse.[3] Despite this innocence, and although contingent effect is not itself

---

[1] Biology, the science of life, has no concept of life (Wolfe, 2010:207).

[2] Meno's Paradox; in Plato's Socratic dialogue Meno, the dilemma how to find something when you don't know what it is that you are looking for.

[3] Presently, it appears that philosophers prefer to argue only with dead biologists and biologists only with dead philosophers (Morange, 2008:xii).

2 Evolution

necessarily indicative of a navigated process, speculative inference has enabled the sampled fossil trail to enliven a comprehensive discipline of hypothesised inquiry wherein the concept of an 'evolution' is employed to refer to both a transformative process and the phylogenic sequence of its living manifestations (Collingwood, 1960:13).

Biology is the study of the living equilibrium, evolution its metamorphosis across time. By contrast, history is a story of the past recaptured from presently available evidence (Collingwood, 1978:96). In biology, that presently available evidence is the found fossil record (Schindewolf, 1993:4). Given, however, that process is serendipitous and the historical archive fragmented, both estimation of credible evolutionary process and the reconstruction of evolution history alike turn upon coincident conflux of unavoided pathways, each the consequence of a once living present wherein an immediate difference made a lasting difference (Bateson, 1972). However, evolution should not be understood merely as the palaeontologist's catalogue of sequenced relics, but rather as trace of an historical-collective dynamic (Ruiz-Mirazo, 2004:323). Ultimately, animate evolution, whether regarded as mechanism or as narrative or as both, is a human rationalisation of the regeneration occasioned when self-referential autopoietic lifeforms establish novel structural and organisational congruence to changed internal or external realities (Escobar, 2012:62).

Systems at equilibrium, whether metabolic or phylogenic, invest in or are invested by inertia, that is, in the conservation of their own organisational properties, any biologically necessary trans-generational survival therefore not of the fittest but of the normal (Uexküll, 1982:62). Accordingly some significant, persistent, or drastic upset is necessary to upset this working symmetry. While it has been reasonably claimed that living organisms are for most of the time self-directed systems, nevertheless they are in all of that time subject to incidental irritation or direct assault. To the extent that this aggravation of a living dynamic may be categorised as extrinsic or intrinsic is relevant only to the relative effect upon the organism of its own obtrusion or resentment. Importantly, in any confronting interaction between discrete intricate systems, each particular change in the state or rate of either independent dynamic is essentially unpredictable, a metaphoric light touch to one participant possibly causing a disproportionate flurry of entailed consequences in the other - and that in the naïve case in which only two such conflicted systems are involved. Many lifeforms, and certainly cells, surrender significant individual autonomy to have advantage of integrative multi-partner unions the which may enjoin just two or, more frequently, millions of self-referential forms into common conserved operational super systems.

Consequently, whilst the serious sciences calculate within predictable Cartesian certainties (but routinely settle for mere probability[1]), evolutionary expression operates within the far more demanding confine of closely delineated but nevertheless unforeseeable possibilities. That is why evolution is more an adventitious process than it is a predictable science, an otherwise necessary demonstrable certainty translated into a convenient if only implicit continuity. This continuity, as common ancestry, is to canonical biology what confidence of antecedent cause is to Newtonian physics (Berlinski, 1996:30) - if likely just as insecurely metaphorical (Lakoff & Johnson, 1999). Arguably, in a world objectively regarded to be without planned purpose (that is, in the method of scientistic materialism), continuity could only ever be an

---

[1] It is suggested by Bekoff that the plural of anecdote is data (2004).

incidental by-product of living, never the positive goal of life (Haukioja, 1982:363). In such a science, evolution happens (Vane-Wright, 2014:230), continuity a happy coincidence not an explicit principle.

To the extent that fossil evidence may be discontinuous is immaterial, for while a punctuated record informs more of phenotype than process (Simpson, 1944:xvii), as proof of a propelled process it is amply sufficient and, importantly, even from this ambiguous perambulation we learn (or reaffirm) one critically significant thing, that life emerges only from life (Bergson, 1998:231); perhaps from a single life (Darwin, 1859).

1.2   Natural kinds

Fixity of species, as opposed to their incremental transformation, became the *cause celebre* of early evolutionary dispute, creationists insistent upon a pristine perfection, the scientific community and their rural allies proclaiming an endemic variability. Anxious, then, to disavow a normative (creationist) attachment to the fixity of species *per* Linnaeus, Darwin set the topic of his signature publication as the origin of species, thereby ensuring that this within-species variation became the advantageous focus of an argument which could furnish sufficient rigour to propel his proposed evolutionary mechanism. In 19th Century biological science, spontaneous variance within populations became the engine of adaptation - an undirected mechanistic means to explain the palpable fit of each species to its accustomed habitat. However, unacknowledged and whether unknown, overlooked, or dismissed by Darwin, a shift 'beyond the Linnaean world of fixed forms and species and into a new world of transformation and evolution' had already been enunciated by Goethe some seven decades before (2009:xxiii) and was still, in Darwin's time, authoritative orthodoxy in Germany.

Variance within a species unquestionably occurs but experience suggests that it is neither as monstrous[1] nor as providential as the Darwinist requires to support eliminative and negative notions of natural selection as the prime cause of animate diversity (Schindewolf, 1993:360). And, of course in any event, 'the precise manner of a character's coming into being is irrelevant

---

[1] As *per* Goldschmidt's postulated 'Hopeful Monsters' (1982). Such monsters presently resurrected (Lima-de-Faria, 2017: 272) as extreme variants periodically emerging 'irrespective of whether the organism likes [*needs*?] them or not'.

to natural selection' (Oyama, 2000:48). Irrelevant perhaps, nevertheless random mutation became central to the subsequent scientific argument as to whether an established species provided the platform for further evolutionary change *per Origin* (variously by selection and drift) or whether, beyond mere local geographical refinement, each established species represented an effectively exhausted strand of elaboration so that initiation of any further significant calibration must be sought elsewhere.[1]

Patently, each individual organism is one of a wider natural kind when association is extended to include habit, organisation, and particularly, morphology (Maturana, 1975:316). Consequently, every living thing we encounter can be ascribed as the exemplification of a particular species, an experientially familiar natural kind adopted *a posteriori* into a formal categorisation systematically designated as the quintessential taxonomic unit (Boyd, 1991:142) and prescribed by a delimiting cluster of necessary and sufficient shared properties. No living, or for that matter extinct, organism can escape being thus administratively

---

[1] Of course, insistence that every lifeform, living, dead or yet to be recovered, is automatically assigned to (or as) a species, means that Darwin was right - but only by the artificial writ of an all-inclusive definition of 'species'.

8   Evolution

assigned into a category of species. In a formal sense, the species is a rather clear-cut unit even if at the margins of observation it may be dependent upon the personal judgement of an examining taxonomist (Goldschmidt, 1933:540).

Strictly, taxonomic segregation into species follows a common human cognitive capacity for object categorisation to clarify observation. Functionally, the conspecific fidelity of a living system is codetermined by its organisation[1] and not solely by the details of structure which realise that particular entity; however, since organisation is seldom directly discernable[2] and because lifeforms interact *via* form (structural coupling), affinities are more usefully recognised by features of shared physical realisation and appearance (Maturana, 2011:501). People of all cultures observe this commonsense notion of natural kinds.[3]

---

[1] Not least, reproductive and domestic habits extravagantly managed to lie within conserved predictable parameters.

[2] Especially in the captured and preserved condition familiar to collector and museum taxonomists.

[3] Weiss (1973:106) warns, however, that categories have only a statistical validity representing the average of a given group of phenomena - allowing systematic order to be accompanied by freedom for individual constituents.

Part One: Evolution 9

Even without resort to a mystical teleology, canalisation into a discrete natural kind provides benefit for an interbreeding cabal of structurally alike individuals to quarantine, by physical occupation or operational hegemony, a particular feeding niche for its own purposes. Effectively, consolidation of an extended family by speciation is evidence of a customising symmetry between group characteristics and a common accessed reality, the two conserved within an equilibrium to which both the reproductively responsive lifeform and a manageably reliable environment variously and continuously contribute. At the species level, effects are biased towards survival of the typical.

Discounting natural selection as an adequate mechanism, and in the sense that speciation and specialisation might be deemed synonymous, then to the extent that an invested characterisation is rendered by changed circumstances to be inopportune, a frustrated (or emancipated) reproductive dynamic may avoid either extinction or a compounding intensification of already challenged properties by conscription, instead, of latent proximate genomic potentials to reconfigure the development of a more resilient, that is a less habitat-restrained, form obtained by regression, retrieval or recombination within its ancestral genome -

the recessive traverse of Schindewolf's invigorated lateral bud (1993). Strictly then, and for this reason, missing links are absent not mislaid, they a fiction of gradualism; 'no teleost fish was ever the ancestor of any amphibian, reptile, bird or mammal' and so no family trees of shared primal ancestors, only present day descendants in common (Hodos & Campbell, 1969:341, 343).

Independent population differentiation, the essence of Darwinian speciation, detracts from any inhered licence to exploit the congenial spread of benefits reliably replenished within a cohesive environ of recurrent resource-creating geographic and climatic parameters. Effectively, if in crude economic terms, the species is equivalent to a franchised cartel both having and taking advantage of a wealth-creating habitat cordoned by the species' own risk-eliminative honed fitness. Accordingly, in the interests of reliability, incumbent species are not plastic on an expansive evolutionary scale but, instead, actively protect by their behaviour both prior investment and present interest in their own particular specialisation.

1.3 Tangible evidence

Because the categorisation of a problem and any subsequent hypothesis to resolve it are intimately enjoined (Uexküll, 1926:ix), it is important to understand that evidence for evolution lies in the fossil record and in nothing else (Schindewolf, 1993:4).[1] Whilst complementary research amply supports a notion of animate evolution, only the unambiguous fossil proves it. Had no fossil ever been preserved it is implausible that the modern conceptualisation of an extended and changing continuity could ever have been responsibly entertained, so robustly defended, or so generously extrapolated.[2]

Erosion of a parental rock and the subsequent relocation and assemblage of its debris by the impacting agencies of wind and water, together with a recurrent raising and lowering of the earth's crust, have delivered, over protracted time, layers of sedimentary rock in which are

---

[1] True, 'Darwin's finches come to us without fossil evidence, only surmise' (Plotkin, 1996.2). Surmise, however, is not highly rated as forensic evidence in any secular jurisdiction.

[2] Observable commonalities of animate form and physiology would, absent fossil evidence, simply appear as a shared essence of living organisms. The discipline of history would be equally problematic absent surviving monument, text or artefact.

entrapped the traces, imprints or relics of contemporaneous lifeforms. Such fossils may also be found entombed in volcanic ash deposits, captured by plant resins or engulfed in bituminous tar flows. These several accidents of preservation provide indelible evidence of the existence of antecedent lifeforms and singularly admit academic inference of archaic phenomena and conditions. Fossils provide evidence in fact, inference made from fossils is testimony to be tested.

Beyond evidential fact lies the tweak of analogy (comparative biology), expert testimony (theoretical logic), and presumption (collegial consensus). Of course, and quite apart from the serendipity of discovery, sampling revealed by the fossil record (even when extensive) is inevitably biased. Whereas facsimile of plant part, of soft bodied animal, impress of microbial mat, trace of burrowing or root penetration may not be uncommon, any potential to fossilisation is itself a selective phenomenon that hugely discriminates in favour of lifeforms already possessing mineralised structures, for example, teeth, armouring or dense skeletal components and especially so for those forms whose habitats lie within or adjacent to areas of flooding, dune drift, or such marine waters as may quickly bury a carcase and so shelter it from scavengers or from a too rapid deterioration. True, subsequent mineralogical modification makes fortuitous survival possible, but other forces have counter effect to erase the record. Older fossils are prone to be overwritten by subsequent geologic process (Raup, 1972:1069) and the most recently deposited sediments disproportionately the first to be exposed to weathering in any crustal uplift (Canfield, 2014:141). Moreover, over many centuries undisciplined human curiosity has led to fossil sites being exhaustively plundered, finds dissipated, and specimens destroyed or separated from context (Mayor, 2011).

This is not occasion to enjoin the several inspirations and disputes of antique geologians or to provide a platitudinous primer in palaeontology beyond that necessary to briefly advise the bases upon which early investigators came to correlate planetary history with fossilised animate remnants. Indeed, a detailed study of the tentative discoveries and personal histories of pioneering geologists might only lend unhelpful substance to historian Sir George Clark's insight that scholarship is a hard core of facts surrounded by a pulp of disputable interpretation (Carr, 2008:8) or, elsewhere and perhaps mischievously, as a dubious core of interpretation surrounded by a pulp of disputable facts (Greene, 1982:292). Neither dubious nor disputable, however, is the instance that discovery of the enormity of geologic time along with the testimony of captured sequentially changing fossil relics rendered the advent of evolutionary theory not only possible but inevitable.

The today normative rule of original horizontality together with the principle of lateral continuity contrived by polymath and cleric Nicolas Steno (1638-1686) to describe sedimentary rock deposition, opened a way to rationalise the temporal layering and physical contortion of crustal formations in such manner as would later be accorded relative chronological credibility by English geologist, engineer and cartographer William Smith (1769-1839). Smith's cardinal contribution was to associate isolate sedimentary strata by the particularity of their included fossils, a technique further refined by German palaeontologist Carl Oppel (1831-1865) who systematised the practice of zone stratigraphy by designation of a reliable categorisation employing discrete index fossils. Despite actively using fossil shells as a determinative constant to align and sequence disparate sedimentary formations, these early geologists apparently did not attribute any biological significance to their correlations. However, growing acceptance of the coincident match of geological progression with a diversifying animate, and especially vertebrate, procession rendered ineluctable the subsequent adoption of an informed evolutionary biology:

> The fossil record contributes uniquely to our understanding the evolutionary processes by tracking the biosphere through deep time on a scale of millions to hundreds of millions of years. The long-term patterns recovered from the Phanerozoic fossil record demonstrate a range of phenomena not obvious from uniformitarian extrapolation, including widespread occurrence of evolutionary stasis/cladogenesis, long-term ecosystem stability, and recurrent intervals of major diversification and mass extinction (Butterfield, 2007:41).

As markers of deep time history, fossils present as immaculate surviving evidence of past form and likely function. However, as guide to process, fossils provide but a coarse filter by which to sift the route of evolutionary development since they ferry scant trace of the essential contributory details of metabolic, replicative or behavioural dynamics.

1.4 Comparative analogy

Before evolutionary theory became a matter of public interest, a discreet disquiet with traditional certitudes, or rather recognition as myth of what before had religiously passed as fact, infected educated thought throughout the broader European culture. Encouraged by this novel social permission (the Enlightenment, Davies, 1997:603), a pragmatic (that is, an Industrial, Riedl, 1983:210) need-to-know interrogation of natural world phenomena inevitably disposed this newly loosed inquiry towards materialist inference, the which within an increasingly technological age naïvely expected of a positivist science the rational retrogressive fit of patent present expression to past, if merely surmised, cause. This inquisitive unrest engendered not so much a clash of scientific and religious sentimentalities as is did the contest between one established (religious) and another emergent (scientific) institution to arrogate for itself the intellectual leadership of societal allegiances (Rupke, 1994:323).

Once the concept of a rational cause had sufficiently matured as to challenge the encultured credo of creation, escalating evidence of ordering within fossiliferous rocks opened the door to educated explanation of mechanism. Even without any substantive appreciation of the geologic timescale involved, contiguous sequences of fossilised lifeforms readily provided plausible description of a temporal procession, but a truly secular explanation demanded that the plausibly more plausible certainty of probability[1] be canvassed by *ex post facto* rationalisation wherein (albeit coached by a little expert advocacy) fossil evidence could be persuaded to tender informing witness to its own formative environment. Accordingly, from a liberal catalogue of eroded ancestral coincidences, significant if still speculative inference of biological, geological, oceanographic and meteorological import was mounted to identify, to explain (or at least authoritatively assert) such past phenomena as lay irretrievably beyond human observation but which might be rendered compassable by astute imagination (Pimental, 2017). Implicit in this extrapolative modelling has been Lyell's (1830-3) retrodictive uniformitarian argument that past geological (and so biological, Greene, 1982:71) events can be known by reference to processes exemplified in the present.

---

[1] These certainties of probability equate to a contrived neglect of difference to generalize and by doing so promote a plausible abstraction Carr, 2008:158).

18 Evolution

Although he energetically opposed any concept of a gradualist evolution, the retrospective comparative method was rigorously administered to (especially vertebrate) fossil interpretation by Georges Cuvier (1769-1832)) within the rule of applying to a reverse-engineered teleological palaeobiology those inductivist, structural, behavioural, and environmental correlations captured by concurrent conceptual or functional 'conditions of existence' (Rudwick, 1997:36) whereby, for example, a fossil species deemed by modern reference to have dentition only suitable for the predator of large prey must not only have lived within an environment where such prey existed but, in their turn, where a sufficiency of these animals themselves would have prospered in a landscape of adequately verdant plant growth, so adding wider implications of munificent temperature and propitious precipitation: each consequential extrapolation equally plausible if less well grounded on the meagre evidence of a sampled tooth or two. While catastrophist Cuvier preferred on religious and observational grounds an evolutionary progression based on extinction followed by renewal, the somewhat less fossil-authenticated theory of natural selection voiced by Darwin (following Lyell) also extrapolated back through geological time by 'causes now in operation' (Penny 2005: 637); a contrast between two present-referenced methods, one of analogous function (final cause), the other a regression of means (first cause).

It may one day prove to have been an overly ingenuous compact, confounding causation with correlation (Weiss, 2012:98), to suppose that lifeforms existing today in fact operate in much the same fashion as did their equivalents or ancestors many millions of years ago despite interventions by the many climatic and terrestrial upheavals we know of and likely the several that we do not (Qin, 2015). Especially so as we continue to discover the extent to which biotic autogenic activity itself has had direct and massive effect upon communities and resources (Jones *et al.*, 1994:373). Of particular challenge to assumptions of retrofit modelling, there are no behavioural fossils (Hodos & Campbell, 1969:343) and it is therefore mere inference, not evidence, to propose that 'simple' organisms must always have performed in similarly behaviourally primitive ways, or that no homoeothermic dinosaur, for example, was possible[1] nor could have been more intelligently driven than, say, the 'more

---

[1] An equally speculative compromise has been proposed that body size can promote Giantothermy, calculated upon volume-insulated retention of generated heat.

advanced' modern marsupial. Yet it is only by such naïve imaginative regression into the past that we can pretend to recreate unrecorded behaviours (modes of living) and therefrom to imagine the congruent environs in which these might have been expressed.

1.5 Expert testimony

By exemplary demonstration of what educated surmise can erect upon a foundation of profound innocence, an eliminative evolutionary obscurantism has progressed by intuition, inference and historic intervention into an objective scientific staple.[1] At first British in essence, natural selection was immediately contested by learned opinion in France, Germany and Russia (Mayr, 2004:41), and, when attachment to Whiggish Darwinism had faltered even in its country of origin, it was embraced by American energies.[2] This last transfer, coming as it did at the conclusion of WWII

---

[1] It has been claimed that fables begin to be current in one generation, are established in the second, become respectable in the third, while in the fourth generation temples are raised in honour of them (Voltaire, *Fragments sur l'Histoire*, article i, in *Euvres*, vol. xxvii, pps 158,159).

[2] The *Society for the Study of Evolution* was formed at St Louis in 1946.

and the unequivocal imposition of US cultural hegemony, rendered all further deference to Continental or Russian agendas superfluous.[1] The history of evolutionary science is, therefore, as much an inventory of provincialism as it is a tale of discovery, the present liberal consensus being that one is either pro- or anti- a stochastically directed Darwinism (Lashin *et al.*, 2012:481). To be caught the wrong side of this defining dualism condemns one as a radical adherent of unscientific animism or betrays an unworthy recidivism into religion.[2] Fortunately, mere consensuality provides no explanation and it is therefore a continuing challenge for Darwin's doctrine of gratuitous mutation and fortuitous selection that an authentic Law-bound science permits no truly random or unpredictable events in nature, only occurrences we do not understand (Prigogine, 1997:155). Random linear change denies march towards equilibrium (Davies, 1995:35).

---

[1] Actually, un-American. Pre-Sputnik communist Russia dismissed *en masse*, and the science of German palaeobiologists Abel, Beurlen, and Schindewolf, deemed tainted as Nazi collaborators (Rieppel, 2012:253). The language of science became American English (Gordin, 2015).

[2] Biological science spiritualises what it disapproves (Hoffmeyer, 2001:382).

Ostensibly, a coherent concept of animate evolution became sensible only at the conjuncture of a structured appreciation of otherwise anomalous fossil forms, the realisation of Earth's extended age, and the relaxation of a hallowed normative perception of the fixity of species. In practice, clerical obstructionism in the courts, in the universities and in parliaments provided an inelastic obstacle to inquiry as many an incautious savant was to discover. Moreover, scientific studies were in the eighteenth and nineteenth centuries conducted in a particularly parochial atmosphere of intensely competing nationalisms, personal antipathies, and partisan or class interests (Rupke, 1994).

In a seminal text, the first authority to employ the term *evolution* with 'any clarity' was Jean Baptiste Lamarck (Simpson, 1961:36). His plausible cogitations, that either use and disuse of organs or that individual experiential encounters with lived circumstances could actively amend an organism's heritable blueprint (*Philosophie Zoologique*, 1809), were at first broadly accepted,[1] subsequently hugely ridiculed, belatedly (if still marginally) rehabilitated (Guerrero-Bosagna, 2012:292), and recently once again actively promoted

---

[1] Indeed, was a commonplace belief of the time (Koonin & Wolf, 2016).

(Jablonka & Lamb, 2014). Of course, the idea of an evolution did not begin with his usage, indeed was already central within the optimising selection principles of Maupertuis (1746), but Lamarck's unequivocal and authoritative publication may be taken as a convenient if arbitrary historical starting point at which time the phenomenon of an evolutionary ascent towards complexity was widely presumed to operate by natural law[1] both as an accumulative process and as the product of adaptive change tending to a level of perfection akin, by analogy, to that of domestic productions (Darwin, 1872 [6th ed.]:xiii-xiv).

To modern critics, Lamarck's anti-creationist credentials are unfairly degraded by his (then ubiquitous pre-Pasteur) attachment to the spontaneous generation of simple lifeforms. Like Newton, Lamarck was an alchemist. And, like Darwin, Lamarck was just guessing (Corning, 2014:243). However, with his guess of the agency of organisms in their own evolution, Lamarck surpassed the intuitions of the next three or more generations of biologists and is only relatively

---

[1] Natural law seen as a 'higher' law and therefore self-evident beyond doubt and ambiguity. In fact 'Nature' may be any agency personified to advance uncritical acceptance.

recently permissibly cited (Barry, 2013) if also still respectably dismissed (Piaget, 1979).

A questionably less philosophical but certainly more popularising argument for a progressive evolution was tendered, at first anonymously, by bookseller Robert Chambers (*Vestiges*, 1844) who declared for animate progress (both structural and mental) over geologic time culminating in Mankind, all according to a pre-ordinate Natural Law. Popular circulation of *Vestiges* gave heretical encouragement to the democratisation of a knowledge which was itself still evolving from native stocks of materialism, affluence, and impiety. With his text, Chambers antagonised both an established clergy and an emergent establishment of scientists by directly engaging with the world at large in a topic and manner hitherto considered the exclusive property of an erudite elite. This effort towards the containment of professional authority remains prevalent today (Berlinski, 1996:26).

Darwinism therefore represents but the culmination of a 19th Century fashion for interpretive progress and institutional posturing. Neither the notion that an evolution had taken place, nor the idea of the survival of the fittest, was new; what was new, was his proposal that one was the mechanism for the other, thereby offering a 'natural' alternative to explanations based on deistic creation (Ho & Fox, 1988:119). In particular, however, Darwin presented his thesis of the conversion of individual variation within interbreeding communities as a linear cause and effect (Lewontin, 1974:4) and did so 'in one single generalisation' (Kropotkin, 1902:1) couched in contemporaneous[1] sociological terms akin to a natural economic process rather than as a procession of developed forms (Cosans, 2009:78). It was, therefore, in this former guise of trade-off and competition that, on Thursday the 24th of November 1859, a formal science of evolution was born with publication of the *The Origin of Species*.[2] Darwin's best-seller was immediately distinguished by the intensity of controversy it engendered (Hull, 1973).

Accident of history casts its own heroes. In particular, and despite insinuation that Charles Darwin contributed very little if anything new to the body of evolutionary speculation, the temerity of his iconic text and the

---

[1] *Origin* was published in the same year as Karl Marx's *Contribution to the Critique of Political Economy*.

[2] Darwin did not himself use the term 'evolution' until 1872 in his sixth and final edition of *Origin*.

controlling (Mazur, 2016:103) cabalistic vigour with which his alleged originality was robustly promoted (Sutton, 2014) acted, in concert, to both eclipse the work of preceding (especially foreign) authors and to sufficiently elevate his standing in a burgeoning scientific establishment as to render it either unnecessary or unwise for subsequent scholars to probe beyond his seminal treatise. In tomorrow's history, Darwin's *Origin* will likely be held notable for the signal effect of his theorising to emplace evolution as the paradigm of biology (Ingold, 1986:2), he providing an immensely important watershed concept that was sufficiently vacuous as to prove infinitely accommodative and motivating, albeit severely hindering the elucidation of actual biologic processes.

Having determined to his own satisfaction the fact of a nontheistic mutability of species as descent with incidental modification rather than one of independent fixed design (Reed, 1978:204), Darwin set himself a narrow task to inductively discover, tangential to the fossil record, a comprehensive principle of tinkering differentiation, the which, when simplistically construed by observer bias to conflate consequence with cause (Maturana & Mpodozis, 2000:278), he then massively generalised to account for every gradation of organic expression, each resting solely upon a discriminative or differentially selective reproductive integrity falling within the open play of chance and necessity (Grene, 1974:185). Darwin's commonsensical metonymic narrative of eliminative survival, whose starting point had been purposeful interference in domestic plant and animal breeding, was thereby prefaced upon manipulation of idiopathic variances within a species; that is, a model of positive individual artificial selection extended by incubated analogy to hatch the universal principle of a negative natural selection. Analogy, of course, is not equivalence, and it has been claimed that this compromised perspective should have alerted him to the fact that selection does not lead to the formation of a new species:

> No matter how morphologically and behaviourally different they become, all dogs remain members of the same species, are capable of interbreeding with other dogs, and will revert to in a few generations to a common feral dog phenotype if allowed to go wild (Shapiro, 2011:121).[1]

---

[1] Darwin's own report (1868) of his experience with numerous varieties but single species of the domestic pigeon should equally have warned Shapiro of Darwin's vivid awareness of this issue (1868). However, Darwin would have undoubtedly distinguished the reversible effect of canine engineering from the dogged intransigence of natural selection.

Whilst variously implying natural selection to be a force, a power, a mechanism or an agency, but mostly depicted as a principle having both positive and negative roles (Escobar, 2012:55), and although himself innocent of both the aetiology of mutation and of inheritance but sensitive to a strict Malthusian competition for intrinsically scarce resources, Darwin's singular genius was to assign utilitarian survivalism a pivotal role as the mechanism of evolutionary phylogenesis. His primary error was to conflate a farmer's calculated investment with the innocent advantage gained by a stud animal,[1] and then to grandiosely extrapolate this irrelevant model of deliberated rural husbandry backwards across Family, Class, Phyla, and Kingdom, as if all were driven by the stockman's surgical cull. As a causal explanation,[2] Darwinian theory conveniently explains everything, evading the commission of any particular event by posit of an evolutionary feed-back loop where cause is certified by consequence. The intellectual leniency of this mechanistic natural selection was not only staunchly supported by an incurious and impious scientific establishment but has since been embraced in

---

[1] A farmyard perspective encouraged by Erkki Haukioja, 1982:371.

[2] Plotkin (1988:7) offers 'Theories are causal explanations'.

its primal form across many other quite disparate disciplines.[1]

Whereas Darwin had earlier sought to invite external affect, August Weismann latterly strove to eliminate it. As had Darwin himself, Weismann considered, equivocated, then cast his own arbitrary principle of somatic perpetuity during that same period of intense intellectual flux when social controversy attended not only every explanation of evolution but at the time when a parallel science of inheritance was slowly emerging from the pioneer insights and experiments of Mendel, Nägeli and others (Bowler, 2009). Encouraged by the progressive elucidation of eukaryote cell structure and replication, the focus of his scientific bravado turned to contest external compatibility as being the prominent engine of structural change in favour of internal random genetic variability unrelated to and isolated from the organism's context (Ho, 1991:339). On publication of Weismann's *The Germ-Plasm* (1892), the primacy of natural selection as evolutionary mechanism was temporarily overshadowed but subsequently subsumed into a hybrid concept of

---

[1] Social (economic) evolution, according to Varela, in common with its biological counterpart, proceeds by doing whatever can be gotten away with.

incidental mutation the which, isolated by an imagined impermeable barrier between somatic and germ cells, permitted organic variation but maintained structural immunity to all environmental influence:

> Weismann's claim that random mutations [*alone*] are the underlying source of creativity in evolution became one of the cornerstones of the nascent science of genetics and ultimately, of a gene-centred evolutionary theory. For a time, Weismann's 'mutation theory' even eclipsed Darwinism (Corning, 2014:243 *clarification in italics added*).

Without doubt the conceptual eclipse occasioned by adoption of Weismann's conjectured barrier helped to overshadow Darwin's own wavering sympathies for Lamarckian variation[1] (Darwin, 1868). Rather fortuitously too, it passed largely unchallenged that a natural selection would in any event be indifferent to the origin of mutation and, because totipotency had been demonstrated by Austrian botanist Gottlieb Haberlandt two years earlier, Weismann's absolutist separation of soma and germ cells was never applicable to plants, protozoans, bacteria, or asexually dividing

---

[1] Pangenesis. A signal difference between Lamarck and Darwin is the pace - Lamarck generational, Darwin gradual, but both accumulative.

animals, that is, to the overwhelming majority of living forms. Nevertheless, by his determined ministry an assertive paradigm of evolution was ushered towards that *cul-de-sac* of reductionist scrutiny within which an emerging professional elite could shun or dismiss all holistic investigation of organisms as mere remnant of an obsolete and amateurish descriptive Natural History or, and worse, of a Germanic Romanticism. Evolution thereafter devolved into a footnote to biochemistry, at least it did so *per* the report of James Watson, co-propounder of the DNA helix (Toulmin, 1992:164).

It was Ronald Fisher's *The Genetical Theory of Natural Selection* (1930) that made of a fashionable capricious mutation and predictable (that is, natural) attrition at once both a respectable science and a flexible model for genetic surmise by recruitment of the abstractive artifice[1] of mathematical probability, he deftly substituting axial coefficients of selection-mediated heritable change within notional clusters (populations) in place of a Darwinian individualistic trial and error erosion; in short, the statistical properties of numbers transmuted into contradistinctive group variables

---

[1] An abstraction is nothing else than the omission of part of the truth, it becomes well founded when the conclusions drawn from it are not vitiated by the omitted truth (Whitehead, 1968:13).

(Corning, 1983:34). Importantly, Fisher's mathematics promoted the incurious faith that stochastic analysis and allegorical abstraction uncover explanatory reality, rather than that they merely interpose a refined but delimiting perspective (Bateson, in Oyama *et al.*, 2001:158). Together, Weismann and Fisher effectively lifted evolutionary biology up from off Darwin's entangled bank onto the laboratory bench, retaining the elegance of an eliminative natural selection to better celebrate the portent of an overarching and intimidating environment as an ineffable principle of nature which posited in advance the solution to everything that is yet to be explained.

Intrusion of categorical prohibition into theory ineluctably creates uncritical dogmatism that inhibits or, and more importantly, may misdirect on-going research (Noble *et al.*, 2014:2237). Apart from the speculative conjoining of micro-evolutionary means to macro-evolutionary outcomes, the tragedy of Theodosius Dobzhansky's foundational gene-centred neo-Darwinist text, *Genetics and the Origin of Species* (1937), was that it served to entrench as pervasive remedy obsolete principles of gradualism, uniformitarianism, and (especially) the monopoly of random process within an obligatory scenario of differential reproduction.[1] A deleterious effect of this 'illiberal and intolerant orthodoxy'[2] was for much subsequent research and human intelligence to be dissipated by peer pressure to defer, in particular, to the paradigmatic agency of natural selection in the same way that earlier naturalists (including Darwin[3]) had been obliged to make token obsequy to a religious omnipotence; in each case an imprinted cultural bias intimidating investigation and (especially) compromising interpretation. Dobzhansky's orthodoxy of a deceptively revisionist neo-Darwinist discourse was subsequently institutionalised by the ardent advocacy and erudite interventions of Ernst Mayer, a prolific author and eloquent apologist for reductionist evolutionary studies.

Without detracting from their many significant insights, the literature reveals that there are as many neo-Darwinist interpretations as there are authors. Clearly,

---

[1] Dobzhansky's celebrated observation that 'the only thing that makes sense in biology is evolution' was effective for resisting creationism in schools, but as biology it is manifestly untrue (Weiss *et al.*, 2011).

[2] Comment by Rudwick, quoted by Schindewolf, cited by Kranich; and here endorsed.

[3] *Origin of Species*, 1859, (1st ed.): 484, referring to 'some one primordial form, into which life was first breathed'.

natural selection has not only unduly absorbed the attention of biologists for generations but has also effectively curtailed insubordinate inquiry by exercise of imperious influence over professional, publishing and teaching institutions. This prejudice remains active today despite the pertinent arguments of those few but valiant early scientists opposed to the tide of selectionist reductionism and whose work is presently trivialised or omitted from education programs.

The disciplinary fear for science, of recidivism into religion, dictates there be an external, preferably implacable and impersonal, agency; the paradigmatic dendritic evolutionary trajectory therefore a product of, firstly, a single mechanistic external driver and, secondly, of endemic but essentially random and therefore arguably reversible change. Random equates to chance, chance a waiver for ignorance (Prigogine, 1979:155), it being the most powerful form of explanation to assert that an issue needs no further explanation. Not, then, to unquestionably support the 'Darwin Industry' is to disturb the ghosts of those Victorian inter- and intra-institutional insecurities gratuitously brought to an uneasy truce by the spectacular success of advancing technologies for which investigative science claims, and is amply afforded, both due and undue credit.

Darwin's effort to promote an anonymous agency of natural selection as the prime mechanism of evolution was a surmise on *par* with aether and phlogiston, it offering a pastoral analogy of plausible progress that initially loosed, but ultimately tightly reined, the ingenuity and integrity of biological research. Since a notion of gradualist evolution was devised (or adopted) to contradict the myth of an instantaneous creation, it is unsurprising that Darwin's own manner of explication was of a kin with that of theology; a blast of examples the validity of each of which depended upon its own premiss, that is, an intellectual construction seeking to satisfy curiosity by a plausible explanation falling beyond actual proof. Accordingly, the relatively modest call begun by Waddington for a further increment to the evolutionary synthesis is today proven inadequate (Lima de-Faria, 1988). Instead, the need is for the all too wounded Darwinian paradigm, with its overarching mechanism, parochial prejudices, and planetary abrasive competition, to be swept aside and a fresh forensic evaluation begun (Noble, 2015:7), not of evolution as coincident gradual process, but as the insinuating enaction of self-realising saltation and opportunistic elaboration.

However, before acquiescing to any new or newer synthesis it is prudent to observe that nothing authenticates natural selection other than the support it receives from its own doctrinaire pundits. In educated circles Darwin's theory is considered sufficiently proven,[1] a banal truth bolstered both by the common consent of self-certifying academics and by the artifice of calculated experimental probabilities. This biased interpretive veneer of Neo-Darwinism has dominated biological science for decades, each concept or metaphor within its burgeoning literature reinforcing the overall mind-set until it is almost impossible to stand outside of it and appreciate how beguiling it is. Indeed, many scientists just do not recognise its fundamentally conceptual nature (Noble, 2015:7).

Fortunately a dissenting scientific enterprise continued and continues to flourish in those countries dismissed by or peripheral to the Atlantic collusion; researchers in Germany, Russia, Scandinavia, South America, in Japan and now in China contributing significant discoveries that are less dependent (except in censored, that is peer

---

[1] If an untested criminal fraud (Sutton, 2014).

reviewed, international journals) upon conventional loyalties and historical connotations.

1.6 Positive persistence

The act of living is demonstrably percontatorial, that is, chronically opportunistic, explorative and, importantly, intelligently so. If then, as is now widely acknowledged, the agency of living organisms has determined the very nature of Earth's surface and atmosphere, it is quite disproportionate to deny life the standing to have also intimately invested in its own constitution. While there is no evidence of a law of necessary biological development (Ingold, 1985:14) there are apodictic physical, chemical and biotic affinities which may, depending upon the dynamics of peculiar interactions, idiosyncratically mesh with prior effect to bias, to impede, or propel fresh elaborative or expansive expression. This rule-bound but exigency-directed elasticity permits inquisitive adoption of novelty whenever constraint is abated by a withdrawn hindrance or by delivery of fresh opportunity within a composite evolutionary trajectory already advantaged or constrained (that is, governed) by antecedent agency, apposite form, or proximate obstacle.

Essentially, then, where not prevented new direction can be seized (Varela *et al.*, 1993:195) such that life can rationally be appraised as positively percontatorial and not merely passively persistent. Even if, however, exuberant elaboration is unattended by intention it is not, of course, absolved from effect. Indeed, elaboration and mutation equally manifest as errant configuration but while elaboration represents elastic seizure of an interactive systemic opportunity, a random mutation derives merely from brittle replicative error. Although these contrasts are significant enough, the substantive conceptual divide lies between conserved effects; incited elaboration leans to potentiate positive fresh structural couplings, chance mutation invites prompt abrasive edit.

The first cause of evolution is life, living organisms fetching their own causes and from which evolution is but the shadow-show[1] of a passing procession, it the captured progressive trace of what happens when a myriad of pliant self-regulating and self-regenerative

---

[1] Bergson's cinematographical frames (1998:336).

conserved living systems[1] idiosyncratically collide or interact with momentum, community, and circumstance. Within such a copious network of interactions there can be no privileged cause (Noble, 2015:7). An evolutionary sequence, then, is not simply record of an external world acting upon life, but of life acting both within and upon an always volatile but sometimes convulsive world. Amidst such a *mêlée*, conforming (viable) characteristics may arise independently of immediate survival prospects.

Importantly, in the intelligent (that is, the living) world, self conservation entails behaviour. Animate volitive behaviour denotes individual and active, moderated but particular selection between available (perceived) options, the which, by subjective inference of attribution and prioritisation, enable an organism (or by the exercise of a common capacity, the group or species) to actively seek advantage or to resile from harm, either disposition positively contributing to its own preservation and possible proliferation. Accordingly, as both reaction to and as agent of change,

---

[1] Living *systems* first postulated by Paul Weiss in 1922 and only relatively recently adopted as a guiding biological principle by, for example, Maturana. Weiss' work not sighted but authoritatively cited by Piaget (1979:55).

animate action commands the central place in evolutionary construction.

Unheralded, tomorrow's biology began between 1926 and 1934 with the publication in Germany of Jacob von Uexküll's accounts of an organism-defined perceptual universe, the *Umwelt*. The books' publication shortly ahead of WWII and the author's political credentials blunted their credibility and delayed translation and distribution in what, post-WWII, developed into a determinedly parochial and reductionist Atlantic establishment. Subsequently, Humberto Maturana reframed Uexküll's imagination of a cognitive insularity within his notion of an autopoietic systems dynamic.[1] As already noted, Maturana's metaphor of an accommodative enactive evolution has as its point of departure a key conceptual shift away from the prescriptive environmental selectionist logic to a proscriptive structurally determinist logic; from the stance wherein everything is forbidden except what is expressly permitted, to that of everything is permitted except what is expressly forbidden - a position

---

[1] Maturana's insights bear a close concord with the antecedent publications of Paul Weiss.

coincidentally adopted in the English Common Law (Megarry, J. [1979] 1 Ch 344:357).[1]

This collateral freedom allows whatever incidental anomaly manages to survive to be conserved, brusque optimisation by opportune random mutation replaced, instead, by accumulation of resident potencies. Whilst random incorporation might or might not gratuitously provide for a felicitous adaptation, positive retention of structural type together with conserved capacities during epigenic diversification allows for that necessary coherence which makes phylogenic self-realisation viable.

1.7 Constrained pathways

The guiding fact of evolution science is the fossil-enabled deduction that all organisms alive today have variously developed in the course of time from strands of similar, if differently elaborated, ancestral forms within a procession of tested structural changes, each significant step in each strand of which has left

---

[1] 'England, is not a country where everything is forbidden except what is expressly permitted: it is a country where everything is permitted except what is expressly forbidden', *Malone v Metropolitan Police Commissioner*.

some paths exhausted, others unexplored and still others incidentally inaccessible. This canalisation has had effect to project a trajectory motivated by a current momentum within constraints set by prior impress, the potentiating vector thereby already schooled by antecedent obstacle and opportunity. In hindsight this contingent catenation can present as a dependent, anterior state-directed incremental progression towards the next. While this may appear the effect, likely process is that the form of a particular realised expression is contained by its antecedents but not determined by them, any emergent elaborative outcome unpredictably eccentric rather than inexorably predictable. Consequently, the emergent form need not express as a realised intermediate species but derive, instead, from a common or periodic pre-expression link on a shared developmental trajectory (Schindewolf, 1993:170; Lima-de-Faria 2017:12), an observation with rational entail that if neither structure nor function are continuous, the only enduring entity is something profoundly unscientific - life itself (Haukioja, 1982:360).

It is professionally impolite for a mere biologist to question the validity of the physicist's Big Bang theory of material inception.¹ Whatever form a pure and infinitely dense virginal energy may have taken, from the outset of the explosion and for some sufficient distance after, there must have been a phase or phases wherein pure energy was atypically expressed as a local intensity or in a particular manifestation and so differentially impressed within an unwritten and so a notionally infinite *mêlée* of emergent radiation, electricity, magnetism, gravity, motion and materialisation. From this developing asymmetry of time and situation, an idiosyncratic evolutionary capture into elementary particles must then have occurred.² Tension between symmetry and asymmetry remains central to the organisational dynamic of matter, and especially so when apparent novel expression only obtains from the increasingly sophisticated recombination of but a few of the already limited fixed combinations that preceded it. Accordingly, these two antithetic states dominate evolutionary phenomena; matter as symmetry and form, energy in asymmetry and

---

¹ Nevertheless, serious doubts do attach to the concept with important implications for the extent and age of the universe and therefore the origin of life and the duration of evolution (Marmet, 1990).

² How homogeneity of particles was thereafter obtained from explosively diverging radial paths is a matter for later scientists to fathom. One facile, since untestable, answer is that it occurred within nanoseconds of the explosion.

function. Although in continuous opposition, as a dynamic dualism whereby each defines the other, form and function were never and cannot now be separated (Lima-de-Faria, 1988:304-5).

In the beginning, then, an irreversible evolutionary pathway of elementary particles forged across unknowable circumstances of volatile momentum and naïve elasticity led to an atomic universe beyond which inchoate but inviolable rule propelled devolving primary matter along ever more obliged channels to define the scope of a subsequent chemical elemental expression. Yet, while the chemist's periodic table charts a vocabulary of but modest dimension, recombination and redirection of molecular effect permit a prodigal if still notionally rigidly structured lexicon of extraordinary facilitative valency to inform the material world. Paradoxically, the subsequent chronic obliquity and extravagance of evolved organic systems are crafted from merely a subset of this already curtailed catalogue of mineral agents, yet evidence the power of an urgent constrained momentum to magnify expeditious effect by interaction between a myriad of peripheral relational embodiments which, themselves, then subtly recombine or are reassigned to robustly modify form and function to expansively reinvest fresh evolutionary parameters (Lima-de-Faria, 1988). Accordingly, while physico-chemical fabrication is limited by definitive narrow restrictions that hold for all time, the intensity of transformations between and within organic species appear of a different dynamic and exotic variability (Lotka, 1945:177), perhaps flowing from the surmised 'chemical imperialism' of living matter (Russell, 1927:27).

Certainly, it would seem impossible to deny that living organisms, lauded on every side as each being unfathomably exquisite and responsive dynamics, should not at several different levels of that living dynamism be able to muster sophisticated means to survive the hammer blows of trans-generational external adversity and periodic internal subversion. For, if all lifeforms are indeed descended from a common ancestor,[1] then no now-living lineage has succumbed to even the most deleterious impost nor failed to take advantage of the most beneficent experiences that this planet has delivered across billions of years. Accordingly, down to the simplest of plants or animalcules (for all must have been equally engaged)

---
[1] The celebrated last common ancestor need not have been a realized individual but instead a point of separation on a shared developmental trajectory (Schindewolf, 1993:170).

the captured potential for latent experiential recombination lies beyond calculation.

Living commits lifeforms to self-maintenance and repair within a congenial, that is an apposite and reliably accessible resource base, confining dissipation by impeding nonconformity. Speciation, the conspicuous exemplar of this biological impedance, conserves operational equilibrium between resource and resourcefulness, consolidating a found or constructed reliability contrived by intricate interplays of habit, structural congruence and organismic competence. Apposite fit of organism within its milieu may be tested by any subsequent critical imbalance sensed by an individual's functioning, the which thereupon variously advances or retards accommodating dynamics to restore congruity. Depending upon the immediacy, severity or setting of a disturbance, animate restorative response may be behavioural, physiological or developmental, ultimately all three. However, for redirected development to occur, the organism's constitution must be either locally overruled or in part redirected from the conservation of a hitherto inured momentum.

Should intrusive effect present as a sufficiently chronic perturbation, organic persistence may induce realignment of zygotic presentation in reproduction, an erstwhile routine replication of the now compromised *baupläne* proving incommensurate with a signalled reality. Overly challenged phenotypic resistance to external shock might well incite genomic portfolio resort to that retrograde typogenetic structural plasticity proposed by Schindewolf (1993) which, analogous to the growing plant should its apical meristem be severed, is obliged to rejuvenate from lower in the stem to discover and test some fresh reconfiguration. In any particular evolutionary cycle, adventitious mutation must prove less readily accessible than advantageous resort to a resident genomic archive:

> Given the nature of genetic change and the retention of these changes in the genome, the likelihood that these changes would become totally erased and the genome returned exactly to an ancestral state is next to nil (Hull,1974:82).

Rather than be erased, redundant characteristics may be repurposed or lie latent so that, while a progression patently occurs, it may be an error to uncritically presume that a more 'primitive' ancestor was rudimentary in the sense of having fewer metabolic processes available to sustain it. Indeed, evidential surmise suggests quite the opposite. Equivalent contemporary bacterial organisms, for example, posses a genetic and biochemical versatility forfeit in the structured complexity of the eukaryote cellular world (Lane 2016:157-8).

1.8 Social context

It is popular belief and an encultured trust that define the horizon of expectation we set for scientific institutions and, therefore, a public (that is, a democratised) science is best couched in those same value terms as are to be found circulating within the conversations of a general community. Pure science, that is one retreated into its own refined objectivity, is akin in terms of social relevance to the distilled theology of an enclosed monastic order. Realising this, and as proponents of an exciting emergent discipline, many Victorian scientists freely engaged with the moral issues of a broad contemporary constituency, indeed so much so it has been claimed that 'the eighteen sixties and seventies saw the triumph of Darwinism over religion, but also the rise of Darwinism as religion' (Grene, 1974:186; Uexküll, 1926:264), with Darwin himself anointed as biology's founding prophet (Shapin, 2010:3-4) in a creed not of perfection but of

Part One: Evolution   49

process, the contingent loss of mankind's supernatural pre-eminence compensated by discovery, instead, of a naturally evolved intellectual pre-eminence (Jackson, 1997:16).

To this newly proselytised laity, evolution as a pervasive secular process became a self-explanatory reality (Toulmin, 1961:40), Darwin's unproven assumptions taught as fact and interrogative argument denied to the extent of a natural selection being absurdly defended as irrefutable by those high priests of the controlling academe whose own relevance depended upon the robust denigration or exclusion of critics (Hayes, 2009:67). Clearly, once an explanatory metaphor becomes normative, argued deviance can be made to present as pathology (Rock, 1973:16) and whatever was actually sought to be explained authoritatively rendered marginal (Webster & Goodwin, 1997:124).

It would, however, be unfair to blame Darwin's acolytes entirely for this situation. It is claimed that all science is a 'socially embedded' religion (Gould, 1994:86) founded upon the 'myth of objectivity' (Lakoff & Johnson, 1980:197) and made viable only by gaining some independent methodological traction. Evidently, and no matter how each may be individually construed or contended, the several subsets of religion and science severally satisfy a common human impulse to frame comprehension (Quine, 1951) within a self-consistent and predictable system (Jones, 1982:100), each a mass of facts ransacked by predilection; in theology to worship obedience (humility), in biology the celebration of survival (wellbeing) (Lewis, 2001:111).

Inevitably, as alike human endeavours, transmutationist science and transcendent religion are as much directed by discourse as by discovery:

> ... the ways in which we talk about scientific objects are not simply determined by empirical evidence but rather actively influence the kind of evidence we seek (and hence are more likely to find) (Keller, 1995:35).

The biblical equivalent, of course, is 'seek, and ye shall find' (Matthew, 7:7). Not merely method may be prejudiced by prior conception, but the very results of empirical research can be impacted in unanticipated ways (Plotkin, 1988:20). In practice, therefore, attachment to a particular explanation can be characterised as stemming not so much from conviction as from prejudice.[1] It, for example, a serious fault of scientific sentiment that it arbitrarily elevates mediaeval rational elegance into a modern methodological sleight: co-opting mischievous scholastic theological parsimony to promote monotheism[2] as a cardinal heuristic with the same ingenuous effect to deflect inquiry, thereby elevating complacency over curiosity in order to deny complex evidential input (Sober, 2008:359). Natural selection is one such engraved simplistic perspective which gravely hampers understanding of transmutational phenomena in what are patently extraordinarily intricate and multifarious interactions of synergies and events.

Both religion and science lay claim to certainty; religious spirituality with its devotion to purpose and its signature disregard for fact standing in stark contrast to a scientific rationalism that holds teleology anathema and fact as holy. In essence, science knowing beyond belief, religion believing beyond knowledge.

---

[1] Or a theory (Carr, 2008:164).

[2] William of Ockham's law of parsimony which, in dispute, shifts the burden of proof towards the most complex of any two competing proposals: the simplistic irreducible power of God's agency thereby overcoming all other contrary explanation. In biology, the heuristic of 'Ockham's razor' is generally applied to issues of selection and systematics as a dubious bias in favour of simple explanation of complex or indefinite problems.

Inevitably, therefore, it is audience predisposition that provides the necessary nuance between a religious or a scientific explanation (Maturana & Varela, 1998:28), one faction aiming to petition 'Nature' the other determined to probe it, both (if differently) having palpable intent to secure some human advantage and each in their own manner striving to capture attention by making the wonderful commonplace - asserting that complexity, correctly viewed, is but mask for simplicity, their art to find incontrovertible pattern hidden in apparent chaos (Simon, 1981:3).

A revisionist study of evolution should be more than the mere dismantling of a doctrinaire Darwinism. However, that doctrine is sufficiently entrenched as it must be challenged before useful study can proceed. Natural selection is not the engine of evolution: in the manner of a creative wand, capricious capacity for mutation can be said to admit every rebellious combination, enduring time permit any fickle effect, and memorising heritability propel every trivial expression. By their mutually inexhaustible interactions these indefinite premises allow plausible process to overawe all further interrogation, the accompanying intrinsic freedom from possibility of validation enabling a broad and inclusive church of universal Darwinism to align with, to appropriate, or itself to be appropriated by otherwise massively incongruent concepts (Levitt *et al.* 2008).

Undoubtedly, to the post-modern relativist the plausibility of imperceptible change extrapolated across a temporarily imperfect fossil trace is sufficiently congenial and resilient that Darwin's proposals continue to unobtrusively present as self-evident fact (Grene, 1974:192). So, too, does the simplistic view of an external homogenous environment dictating, by the negative eliminative principle of natural selection, the direction of all animate form and function. By contrast, it is increasingly held by many that discrete living organisms 'are systems such that nothing external to them can specify what happens in them, and that any external agent impinging on them can only trigger in them structural changes determined in their structure' (Maturana & Mpodozis 2000:264). Structural change, selected or systematic, entails fresh physiological and behavioural eccentricities.

In the event, Darwinism has long been (Sedgwick,1859) and must still be reckoned unscientific (Popper 1976) in any Baconian sense,[1] it relying on assimilative theorising that is invincible by either falsification or observation (Ho & Saunders, 1979:574, 574) and, more importantly, eliminates that touchstone of science, predictability. It is therefore unsurprising that a spirited and discursive but irreconcilable controversy has long decorated the profession of biological science, drawing into its discourse no few chemists, physicists, philosophers and theologians eager to conjure resolute conclusions from their own or others' speculative theorising. Fortunately, many an interesting discovery may be rescued from fallacious premiss and, certainly, Darwin's sweeping evolutionary generalisms have brushed scholarly inquiry into several dark spaces that would not otherwise have been so self-reassuringly explored. In particular, Darwin's genealogical tree places us reassuringly atop the animal kingdom; unique if not uniquely so (Griffin, 2001:253).

While much multidisciplinary theory firmly holds to a first edition Darwinian doctrine, a profusion of respectable heresies (Grene, 1974:191) and improbable

---

[1] The Baconian scientific method does not take an investigator very far into the reality of the living organism (Grene, 1974:239). Largely, perhaps, because the 'authority' of scientific ethos is now applied to areas of which the founders of the methodology knew nothing (Hobhouse,1926).

extrapolations collude to suggest that Darwin's own thoughts need only be sparingly cited as an ingratiating endorsement. It is still useful, however, to reflect upon the implicit bias and essential weakness of his rule of evidence:

> If it could be demonstrated that any complex organ existed, which could not possibly have been formed by numerous, successive, slight modifications, my theory would absolutely break down (Darwin, 2014:189).

Of course, the contrary proof of finding any complex organ that had indeed been unambiguously formed by numerous, successive and slight modifications is safely and equally beyond demonstration. Consequently, and with the hindsighted vision of all miracles, optimal adaptation by gradual natural selection can be claimed to create anything. Indeed, it is said that 'the words *natural selection* play a role in the vocabulary of the evolutionary biologist similar to the word *God* in ordinary language' (Gatlin, 1972:164 *italics in original*). Such is the temper of a socially sensitive science.[1]

---

[1] Bertrand Russell suggests that evolution 'is not a truly scientific philosophy, either in its method or in the problems which it considers' (2009:8).

Part Two: Life

2.1 Elusive origin

We being unrepresentative models of general biological principles (Goldenfeld, 2016:348), our own act of living affords little privileged insight as to the nature of life but, rather, frequently interposes unfordable bias. Each of us may, if without inviting consensus, describe what life appears to be for our individual selves while projecting this characterisation onto others by pretending to infer its diffused effect in ever more different lifeforms until we reach the point of wondering what, if anything, we share with those miniscule or sessile biota into which life apparently infuses scant realities of purpose or awareness. That commonality, no matter how modest or complex the lifeform, is a style of organisation. It is a particular mesh of organisational association, not just a cluster of parental genes, that each organism personifies and conserves.

A pastiche of educated surmise culled from the current literature posits life as the loci of a spontaneous entropy-denying convoluted ecology of autocatalytic abiotic composomes contained within their own boundary and set upon an adventitious spiral of communal self-promotion: in essence, a self-assembling and intensifying hypercycle of interactive replicating organic macromolecules within which time, rule, circumstance or affinity, either conduct or permit cofactors first to cluster, then to compound, to recruit, to concentrate, to measure and finally to reliably sequence interactivity as to synchronise these recombinant reactions into a spatially discrete, determinatively self-templating, memorised and managed cascade of context-dependent hierarchical (that is, interactive, interdependent, cooperative or resonate) constraints by which a distributed self-referring and self-generating holistic matrix systemically contracts, and is able to project and protect, a homeostatic, pedigreed, and aware (that is, a self-conscious) autonomy.[1]

---

[1] Aquinas' almost secular 'matter articulate'. Many employments of words such as 'force', 'vitality' and 'organization' are deployed in contexts akin to the less welcome term 'soul'.

Thus eruditely uninformed, it needs be noted that life has also been variously claimed to be configured as an inherent tendency (Maupertuis, 1745), a natural drive (Bergson, 1911), an irreducible mode of being (Compton, 1984), as self organisation (Kauffman, 1996), in terms of a compounding complexity (Casti, 1994), an inured elemental simplicity (Lima-de-Faria, 1983), or as a nested system (Maturana & Varela, 1980), each configuration (as Darwin discerned) becoming increasing efficient for no better reason (if reason there be) than to survive[1] - for what other benefit could life provide for itself other than the persistence of its own living.

The fact that spontaneous order can emerge in simple physical and chemical systems is widely recognised. However, an extended proclivity for systematic self-ordering finds cardinal effect in plant and animal, all obliged physicochemical conformation conspicuously metamorphosing into an obliging self-regulation. Accordingly, and by richly diverse means, each organism actively and diligently attends to the service of its own metabolism by engaging competencies and

---
[1] Whether that efficiency be directed as shaping to a stable environment or as urgent response to a world in confusion has been a matter of argument, geological history suggesting periods of both pronounced stasis and urgent catastrophe.

symbionts (Sapp, 1994), by exploiting external resources, by dedication to self-maintenance, and by election of an appropriate estimate of immediate opportunity or prescience of devolving challenge. Intraindividual self-direction importunes self-management, a conceptual terminology encompassing regulatory processes preferred to the more commonly employed Kantian notion of self-organisation for, whereas even a trivial ordering may secure predictable regularity, management intones directive agencies of coherent participation by both anticipatory and correctional intervention. The problematic biological issue, of course, is not the fact of these phenomena but their origin.

Incomprehensible time and absent proof grant free rein to scientistic[1] speculation, admitting opportunity for authoritative if unsecured conviction to respectably stand in the place of explanation. Yet, even should we stumble, or indeed have already stumbled, upon a sufficient interpretation of life's origin, its intricacy (or simplicity, Maturana, 1975:325) may well be beyond our power or prejudice to grasp and the vain quest be

---

[1] Scientistic rather than scientific because 'science ignores time *via* objectivity and exactitude, in order to escape into a lasting present (Grene, 1974:251). Time is a crucial actor in evolutionary studies.

obliged to blindly continue (Jeans, 1981:10). Despite it being counterintuitive 'to suppose that crude matter, obeying mechanical laws, was originally its own architect' (Meredith, 1928:lxxvii), Victorian scientists colluded to agree that life must be capable of a material explanation (Minot, 1902:2), the responsive skein of self-referencing organisation but a single enervating fibre differently woven across the millions of plant and animal lifeforms, and that this energy-shepherding, entropy-resisting strand be self-perpetuating, resilient, opportunistic and universally valuable enough to be strenuously promoted or protected by all the transient and various forms charged to weave it.

Even if indeed spun from a single thread, ethereal life and corporeal living could (or should) trace discrete trajectories, one in the determined conservation of a universal persistent core dynamic, the other towards a tenuous local congruence propelled by the continuous interplay of operation and opportunity. Certainly incidental, because elaborate diversification and intricate adaptation would only positively ensure life's continuity were its early and so more primitive forms be endangered by their own simplicity: but this does not not appear to be the case. 'Simple' lifeforms are conspicuously the most successful survivors and life-carriers. However, and against any further simplistic posit of simplicity, prokaryotic sophistication may now be acknowledged to the extent that gone are 'the times when the bacterial world was considered as the simplest biological kingdom, devoid of complexity, and a remnant of the primitive manifestations of life' (Marijuán et al., 2010:100). In evolutionary terms, it is far too late for any protean innocence to have survived.

Essentially, of course, whether found by revelation or by research, it is rationally possible and therefore reasonably probable[1] that two or more discrete and parallel 'origins' of life have occurred (Gould, 1994). In fact, if two, why not fifty[2] (Buckmaster, 1932); inference from surmise is limited only by audacity. Darwin suggested ten (1859:484). Of course, there may not be any unitary 'life' left for us to uncover but, instead, many derived strands of living the inceptive dynamics of which were long ago[3] radiated into trajectories of discrete divergent manifestations leaving

---

[1] There is no calculation for the probability of a single nonrecurring event.

[2] UK Law Lord's goading argument against incautious escalation, AC562:578.

[3] In particular Cambrian prototypical animal forms and Silurian land plants.

behind a barren pivot around which today subsists a peripheral palette of clustered lifeforms (Martin, 2011) each of which is still open to some form of dendritic expansion and so progressively ever more mutually incapable of lateral recombination.[1] Every distinctive strand of 'life' must, therefore, initially have been a singular trick of nested tensions that is now too developmentally dissipated and refined to be re-conjured or parsed outside of the insulated dynamic of its present proximate milieu.

Should the origin of life be recognised as a unique phenomenon independent of any antecedent devolution and so accounted a legitimate scientific research project, biologists should first attend to the core premiss whether life is now or ever was a single entity (be it common process or dynamic) or at the very least to interrogate the origin of such an audacious presumption of solitary cause that speaks so eloquently of vitalism; for, if to the contrary, life were to prove multifarious, then we could reasonably expect disparate origins and not a linear concatenation of evolved organisms (life's disposable husks). Indeed, we might discover an underlying metamorphosis of life itself - not merely as a

---

[1] That is, at the organism level: not necessarily so at the cellular componentry scale.

contingent transmutation of form but as the evolution of process.[1] In short, it would be reassuring to know what evolves, life or living.

Self suggests agency. The lifeform's consciousness of its own selfhood together with an associated compulsion to satisfy selfish drives is claimed as the fundament of life and living (Vane-Wright, 2014:223) and that it is by this recital as an operational self-referential system that organisms are considered intelligent, the very concept of a self itself the signal threshold:

> Not just animals are conscious but every organized being is conscious. In the simplest sense, consciousness is an awareness (has knowledge) of the outside world' (Margulis & Sagan, 1995:122).

Whether it be a unicellular or a more complex lifeform, organisms do things based on knowledge of what is happening in and about themselves (Mazur, 2016:44). Therefore, and no matter how rudimentary its inaugural

---

[1] Carl Woese (Mazur, 2016 :343) maintained that evolution is not a procession of forms but rather the procession of processes, that life changes and accommodates newly empowered acts of living. Woese was mainly concerned with pre-cellular self-referential endosymbiotic processes (355).

expression, life as we know it (or as we prefer to envision it) began when an organic entity within a progenote cluster registered a barrier to otherness, an immediate organismic alienation introduced to the degree by which a now quintessential self was aware of other intersecting selves (externalities) and thereby of a consequent imperative to accommodate, evade, or to exploit those others' apparent ambitions (Godfrey-Smith, 2017:36). 'The adaptive relationship with the environment is a *sine qua non* condition for any intelligent system' (Marijuán *et al.*, 2010); what likely developed as a facility to better manoeuvre within the presenting milieu has unarguably and increasingly evolved into a power to manipulate it.[1]

Ultimately, a satisfactory origin of life as we know it may well remain a matter of axiomatic concord or statistical modelling rather than of any empirical science. However, given the expansive contemplations of physicists, the reductive curiosities of (bio)chemists, and the more than sufficient challenge of accessible contemporary research topics, it is perhaps not unrealistic (if still unduly timid) for practising biologists to evade the intricacies of the origin of life and to, instead, pursue a schooled disciplinary profession in

---

[1] Waddington's 'exploitive systems'.

exactly the same manner as have their theological counterparts, that is, to construct a cohesive whole within a paradigm beyond remit of verifiable elucidation. Ask any parish priest about the origin of God.

2.2   Initial process

Scientific respectability insists upon a creationist explosive start to the universe and so of all to be found within it. Life, therefore, must be an entail of this event. However, and with regard to assumptions contingent upon any such a lively galactic detonation, Popper (1991:143) prudently cautions that 'the precariousness of the status of this speculation need not be stressed'.[1] Still, if it could be proven, rather than merely inferred, that at this birth of entropy (Greene, 2005) the mass of the Earth plus that of the solar system or of the entire universe of galaxies arose from a pinpoint of infinite density, then merely demonstrating the origin of life as a subsequent if incidental trick of

---

[1] The inevitable realization that the Big Bang theory of creation is ultimately irremediably flawed must, as with the displacement of Natural Selection as prime evolutionary mechanism, render much (practically all) scientific theorising to be inadequate or misdirected, see Marmet (1990).

combinatory chemistry must surely pose as little more than a trifle. Unfortunately, no such trifle has as yet unambiguously presented itself.

Whatever universe-creating scenario one may choose to adopt, it is important to remember that life emerged from and was governed by the symmetries of a rigid mineral world which must have itself devolved both physically and chemically (Lima-de-Faria, 1983) in an evolutionary process that preceded, indeed generated, genes and heredity as we now envisage them. However, this inorganic distillation of energy, no matter how idiosyncratically arrived at, seems to have consolidated to such a sufficient degree as few scientists presently expect further particles or elements to be evolved rather than merely contrived or conceptually discovered.[1] By contrast, and outside of any antique caste of fixed species adherents or homocentric acolytes of mankind as the supreme and ultimate being, future organic evolution is regarded as inevitable. It would therefore seem that pre-life evolution was driven by discharge of a constitutive capacitance whereas lifeforms emerge from and continue to be elaborated by convoluted interaction with volatile exigencies. The essential if

---

[1] Mendeleev's Periodic Table was premised upon the assumption that the laws of chemistry were *absolutely* constant.

subtle difference is that whilst inorganic matter changes, only organic life morphs (Bennett, 2010:51).

Generally, aside from the disingenuous premiss that the primary requirement for the origin of life is the absence of life (Levins & Lewontin, 1985:46) or, more sensibly, that living organisms must exist and reproduce before they can contribute to an evolutionary network (Escobar, 2012:56), most scientists propose that life emerged under circumstances inhospitable to life, at least and more securely, to life as we know it. It would appear that upon the surface of a planet where emergent life was the creature of a compelled clever chemistry, near-life did not long survive the living. Either the hazards of a lived-in environment or the hazardous environ in which only life could live, proved sufficient to disrupt an ancient, previously persistent, and erstwhile burgeoning prebiotic or protobiotic dynamic. Strictures attending any transient co-existence patently proved inimical to one and decisive for both. Paradoxically, it was not the vulnerable nascent lifeform but a venerable prebiotic chemistry that succumbed to this novel and persistent juxtaposition. Subsequently, living organisms have proven to posses a conspicuously robust and felicitous plasticity, proactively spreading around the planet to invade every transient crevice whilst surviving both predictable local depredation and capricious global catastrophe. Whereas each lifeform is structurally contained by its own reality, living has proven an incredibly flexible functional elaboration. Evolution is testament to that versatility.

We are now encouraged to believe that the hazardous environ in which only life could live contained free molecular oxygen (Canfield, 2014), a gas poisonous to all (most) forms of life as we know it in the absence of suitable mediating enzymes (Cloud, 1972:544). Although this atmospheric novelty may well have had definitive effect upon multicellular development (Gould, 1994:88), it clearly would have had greater effect still to permit land invasion, to enable structural complexity and to promote body size (Ward & Kirschvink, 2015). However, if the oxygen concentration was produced by living things in the first place then, clearly, it was hardly unfamiliar or peculiarly toxic and therefore some complementary change must have preceded or accompanied it to trigger such exuberant animate diversification. Interestingly, the profligacy of Carboniferous plant life has been floated as the source of this free oxygen (Bidwell, 1979:191), the associatively sequestered carbon now being diligently exchanged into the atmosphere by our equally profligate combustion of fossil fuels.

The conventional depiction of life as the fragile quirk of a combinatory chemistry beset by an intimidating and convulsive externality whilst all the while sapped by appetite and senescence, invites the conclusion that living has perilously survived change across planetary time only by virtue of the incidental promotion, or sifted retention, of its own unintended deformities. External change fortuitously favouring a gratuitously heritable replicative error: Darwinism.

Equally, but no less arbitrarily, should instead of a theorised vulnerability life be accounted to own the reality of its recorded tenacious persistence and sweeping planetary effects, then passive shaping must give way to active insinuation, initiative outpacing selection, participation ousting mutation: Vitalism.

It is, however, a grave if common error to insist upon this simple dichotomy of either Vitalism or Darwinism (Lima-de-Faria, 1983:1083), that is, between mysticism or fear of mysticism. There may be many other theoretical possibilities; either turn, life remains as real as gravity[1] (Thomson, 1871:269) but while gravity has no history, life does (Hull, 1974:86).

Disciplined logical contention requires proffer of an inaugural implicit or irrefutable fiction (supposition or premiss), from which further plausible investigator-inspired conceptualisation may proceed (Uexküll, 1926:ix). Formal science instructs a creationist Big Bang start to our universe, a multidirectional explosion of symmetrical properties diverging at once from the point of infinite density into an asymmetric flux of materiality and entropy. The plausibility of this expedient origin thus comfortably entertained, an evolution of energy into matter at first required formation of elementary particles, those inferred units of universal construction which by their opportune or necessary gravitations determined or directed a further evolution, that of the chemical elements. This evolution of the constant elements, already canalised by the atomic characteristics of its constituent particles, autonomously followed such available trajectories as encountered chance or intrinsic valence dictated with effect to yet further prejudice the matrices of a subsequent third evolution, that of minerals. From this

---

[1] If not, perhaps, as 'universal' in the Newtonian sense. Darwin found gravity a suitable metaphor in closing his account of *Origin*.

already thrice-distilled primordial energy, now latterly presenting in a tangible world of minerals, life emerged on planet Earth as a fourth convolution, it similarly but further subject to all the funnelling effects of prior combinations and constraints. In essence, while the hypothesised primordial creative explosion of energy led to a massive dispersion of matter in space, consolidation of that matter paradoxically conscripted opposite coalescent effect, any notionally infinite range of possible expressions accumulatively curtailed as each investigated pathway became satiated; such that that, from a consequently modest alphabet of chemical elements elaborately displayed in minerals, planetary life was obliged by energetic necessitude to pursue the expansive dynamic of its own founding energy by creative writ of a functional complexity, each new intricacy inviting an ever greater inter-relational sophistication to sit within prior circumscribed but freshly relevant parameters (Lima-de-Faria, 1988).

If, as is now sufficiently understood, the overwhelming numerical majority of living organisms[1] are unicellular and that even the most complex of lifeforms are composed of cells (Schwann, 1847; Wolpert, 1995;

---

[1] A view that reflects a conventional if dubious position that viruses are not living forms.

Capra, 2002) which are themselves constructed from cells by cells[1] (Virchow's essentially vitalistic *omnis cellula e cellula*), then biology as we know it credibly begins[2] with and effectively equates to the singularly discontinuous[3] emergence of the atomistic undifferentiated cell (Villarreall, 2005:101) envisaged by Maturana (2011) as a first order molecular interactive autopoietic system.[4] Such few opponents as there are to this collusion either argue peripheral morphological, physiological and embryonic exceptions (Reynolds, 2010) or seek a radical revision of the isolate cell doctrine (Goodwin, 1963) for at least, but not only, supracellular plant life (Baluska *et al*. 2004).

That Cell Theory and Evolution Theory, key generalisations of the contemporary biological paradigm (Mazzarello, 1999:E15), alike find expedient common resort to a problematic prebiotic

---

[1] Canguilhem, 2008:46, '... the obstacles to the omnivalence of cell theory are almost as considerable as the facts it is asked to explain.' For a referenced critique of cell theory refer pps 25-56.

[2] Stuart Newman proposes that the cell is too sophisticated and, especially, too biologically comprehensive to have been the first lifeform (Mazur, 2015).

[3] That is, involving lateral as well as linear exchange of DNA and other cell componentry.

[4] Of course, the cell may not be a 'thing in and of itself 'as much as a nexus of interactions.

physicochemical spontaneity[1] need not surprise us. Indeed, much scientific theory calls upon the luxuriance of facilitative miracle[2] presented as an innocuous commonplace (McKenna, cited Sheldrake, 2009:3) whenever the savant requires to affect an ingenuous transposition of description into prescription by which to extend a plausible narrative or workable axiom, it invisibly serving as an accommodative fiction (Vaihinger, 1935). Note that 'both Darwin and Newton, the two giants of English science, advanced theories that relied upon God to perform at least one miracle after the first moment of creation' (Cosans, 2009:103); one, an invisible hand to rewind the celestial timepiece from time to time, the other, to strike one or more primordial life-creating sparks:

> This introduction of an unexplained elementary factor is by no means peculiar to biology, and there should be no objection to it being unscientific. Completely analogous is the use in the discipline of advanced theoretical physics of the terms force fields, elemental quanta, energy, and so on, graphic concepts and descriptions of

---

[1] Spontaneity in a rhetorical sense perhaps, since it is estimated that the eukaryote cell took 1 or 2 billion years to evolve (Noble, 2008:3004).

[2] A miracle, as effect without a cause.

phenomena whose actual essence and real cause are not known. What we do know is only that there *are* such things as field effects and gravity and how they manifest themselves, but not *what* they actually are, *why* they exist, and *whence* they come (Schindewolf, 1993:373 *italics in original*).

Human cognition encompasses a totality which includes both apparent *phenomena*, that is, those realities accessible to sensory observation, and also such conceptually necessary *noumena* as fall beyond perception but are cogently present by virtue of their manifest agency. Note:

> *Sensibility* is the *receptivity* of a subject by which it is possible for its representative state to be affected in a certain way by the presence of some object. *Intelligence*, rationality, is the *faculty* of a subject by which it is able to represent to itself what by its quality cannot enter the senses (Kant, 1894:50 *italics in original*).

Falling beyond human perception does not make a noumenon unreal in any absolute sense since every taxon-specific reality is discriminately determined. In biological discourse 'life' is a conceptual representation (in the Kantian sense), natural selection, hierarchical organisation, autopoietic systems, and indeed Nature itself are similarly naïve but conceptually convenient concepts which may, in the Victorian Monist tradition, be sufficiently explained by assuming they require no explanation (Minot, 1902:3).

Living things are an extension of and retain features of the minerals evolution (Lima-de-Faria, 2017:264). In that mineral world any concept of time is meaningless (Schindewolf, 1993:176), yet time is author of a conspicuous distinction between the merely ordered physicochemical manifestation and the volatile dynamism of a biological organisation (Schützenberger, cited Berlinski, 1996:39). Nothing living is timeless, indeed, time sensitivity is a universal feature of cellular organization, it said to be extremely difficult to tamper with the timer without killing the cell (Goodwin, 1963:5). It might reasonably be ventured, therefore, that proto-life may have found foundation in a fusion of time and physicality, the ensuing temporality becoming radically re-dimensioned to capture and propel a cyclic periodicity towards a responsive linear continuity. Tensioned within this molecularly determined alloy, physics supports form, chemistry attunes function, while time adds sequence as animation; the hidden variable of a tethered *kairos* lending antecedent causal force to maintain momentum:

> Figuratively speaking, matter at equilibrium, with no arrow of time, is "blind", but with the arrow of time, it begins to "see". Without this new coherence due to irreversible, nonequilibrium processes, life on earth would be impossible to envision. The claim that the arrow of time is "only phenomenological", or subjective, is therefore absurd. We are actually the children of the arrow of time, of evolution, not its progenitors (Prigogine,1997:3).

Cyclic periodicity would have proven especially opportune in a frigid prebiotic world by the recurring effect of a planetary orbit to potentiate a tension between the mediating alternation of radiant energy and its deprivation, a reiterative rotation serving to alternately expedite and retard (that is, to pattern) chemical reaction. Thus primed, an elaborative ratchet of resonant molecular combination and recombination could have enabled escalating levels of connectivity and higher-order systemic integrity within a newly engaged linear time (Prigogine, 1997:73); critically, to propel this into an entity exerting independent (that is, a subjective) momentum. The life-promising moment, then, when imposed rotary interval spun off into an autonomous longitudinal sequence. Importantly for the particularity of this living transition, and whereas the autonomy of the holistic lifeform depends on fluid external relationships within an accessible milieu:

> The most significant aspect of self-assembly from the point of view of molecular and chromosomal evolution is that the ordered aggregation of molecules, macromolecules and supramolecular systems takes place independently of external information (Lima-de-Faria, 1988:117).

To what extent, then, rehearsal of temporal periodicity remained an external informant after having evocative[1] effect may not be discovered. However, within the alloy of a manipulative marriage of molecular time to matter, already generated rhythms could have acted as a society of interacting agents and, now independent of the cyclic prompt, conserved and managed emergent oscillations to aggregately compel facilitation of that '*la fixité du milieu intérieur*' proclaimed by Claude Bernard (Gross, 1998) or, as reframed by Maturana (Maturana & Varela, 1980) to maintain the integrity (equilibrium) of a systemic autopoietic metabolic autonomy,[2] the primordial clock lending not only power to correlate a

---

[1] Waddington's choice of vocabulary to describe a non-specific stimulus that initiates a process already but independently poised to unfold.

[2] Conservation does not imply conservatism (Harris, 1997).

linear sequence of events but, and on a longer scale, to pace trans-seasonal as well as trans-generational developments independent of anomalous immediate environmental clues (Bidwell, 1979:472). Understandably, twenty-four hour cycles are virtually universal in biology, a generic aspect of physiology in organisms ranging from algae to whales (Solé & Goodwin, 2000:95).

If, after all, life really is as real as gravity and a directional gravity can be relatively dismissed as the intrinsic property (that is, an uncaused inhered trick[1]) of mass in space, then surely an ordered procession of life can reasonably be discovered as the property, not just of combinatory chemistry, but as an irreversible trick of time in matter.

Of course, whether equated condition or consequence, it is still as much surmise as explanation to propose that life is the creature of time out of chemistry, of sequential orderings set one against another, or that it be hierarchically layered spatially and temporally to provide a structured organisation itself further

---

[1] Within, that is, the modern mythology of a revisionist rationale of science which reinterprets Newton who, himself, insisted gravity be an immaterial force of the universe and not a property of matter (Jacob & Jacob, 1980:265).

compounded into a self-replicating supersystem. Indeed, whenever a commentator realises the impossibility of further cogent elucidation, resort may be had to such a nested complexity of interacting systems.[1] If only for the purpose of illustrative analogy, then, lifeforms have been routinely described as structurally contained operational systems interacting with yet other conserved systems; each organism itself a system of lesser systems which in unison contribute to a whole, the which is, again, part of a wider system acting synergistically with other living and non-living systems. Such intricate interactive networks are conceived as functional associations not architectures, are metaphor rather than paradigm (Pisani, 2007, reporting Fritjof Capra), their near-infinite networks, generic organisations and autonomous systems contrived mechanistic imageries of what others have more economically envisaged as an *élan vital*.

Just as industrial invention inspired the 19th Century inquirer so, today, a pervasive relativity adds systemic complexity to discovery; living systems becoming conserved but accommodative forces, neatly reduced to

---

[1] Design (adaptation), order (organization) and harmony (ecology) were, before their adoption by science, held to be patent manifestations of deistic mercy in the universe (Jacob & Jacob, 1980:258).

mechanical physics in order to prevent a puzzling event being otherwise unhelpfully vitalised. There is, of course, a profound conceptual difference, if of little practical moment, between an organised mechanical force and an irreducible vital force. In either event, patterned process points to anticipated end and so recruits purpose (Grene, 1974:201) requiring management. Because lifeforms uniquely conscript the effects of active interaction with other autonomous or independent externalities, evolving organisms have developed an ever expanding capacity to reach beyond, or manage, their inaugural physicochemical limitations. Intelligent behaviour is the conspicuous token of such a burgeoning capacity. It is, however, important to understand that this, or any, extended facility derives from novel recombination rather than a corruptive mutation (Lima-de-Faria, 1988:203).

2.3 Heredity

Heredity in reproduction, the grease of animate evolution theory, instructs an invigorative path between generations that enables periodic evasion of wear and tear as well as relief from senescence and obsolescence, it buffering each lineage of lifeforms from those unremitting persecutions of external impost and internal corruption for which 'an infinite duration would present a very inopportune luxury' (Jacob, 1982:50). Whilst metaphors of inheritance can mislead (Webster & Goodwin, 1997:124), between these intergenerational reinfections it is retention, not novelty, that ensures coherence; proven structure sustaining a central viability whilst interfacing competencies secure the economies of a local congruence. Heritability admits of no rigid rule nor does it equate to inevitability (Gould, 1987:186) but, as a marginally malleable molecular mechanism acting through already expressed lifeforms, it promotes continuity by providing means to both conserve and modify; a necessary tension curated by transferable intergenerational memory having signal effect to recycle the lived experience and thereby transmit an informed discretion to impose or relax (that is, to regulate) chromosomal replicative templates.

Understandably within the imprecise scientific discipline of a biology paradoxically predicated upon continuity but exemplified by paradigmatic random dispersion, the mechanism of heredity in reproduction was ever a topic of curiosity and speculation. Galton's centripetal regression to the mean, Darwin's centrifugal speciation, Weismann's impermeable somatic barrier,

Mendel's particulate traits, and Fisher's abstracted trends, each served by either adoption, rejection, or amalgamation, to usher a post-Victorian era of gene supremacy to which discovery of the DNA double helix afforded a triumphant reductionist flourish (Woese, 2004:175).

Today, the reified gene (caretaker of form and function, Lima-de-Faria, 1997:119) is depersonalised as simply a particular stretch within a discrete DNA sequence, it both creature and captive of the totipotent zygote[1], inert until activated and so never more than a contingent facilitative agent (Noble, 2015:9), it as fluid[2] and changeable as the rest of the organism (Webster & Goodwin, 1997:251). Moreover, DNA and its likely precursor RNA are presently *passé* as prime determinatives (Luisi, 2016), instead the generously inclusive and elastic genome is widely upheld as heredity incarnate (Weiss *et al.*, 2011), more a database for the reconstitution of successful organisms than as a program to create them (Noble, 2008 a:xii). It is useful

---

[1] Of ovum and sperm - 'Once they fuse, these two highly specialised cells form one cell which is so unspecialised it is totipotent and gives rise to every cell in the human body, and the placenta. This is the zygote, at the very top of Waddington's epigenetic landscape' (Carey, 2012:121)

[2] Different genes do different things differently.

to carefully reflect that in terms of the initiation and fundamental direction of life, the decisive and significant events were those which preceded DNA (Lima-de-Faria, 1983:1065).

Paradoxically, the genome is at once an intractable guardian of the *baupläne*, processor rather than programmer, a discretionary door through which epigenetic influence may be admitted or, in extreme cases, a repository from within which any compulsion to drastic reversion can rummage for retrievable congruities. That the one and the same component of a cell is responsible for both rude caterpillar and exotic butterfly is testimony to the amplitude of this attribution. Essentially, of course, that such a broad scope of mutually incompatible critical functions is attributed to a less than tangible ideation not dissimilar to the vitalist's life force (Oyama, 2000) is itself evidence of our continuing innocence of, but conceptual reliance upon, the mosaics of hereditable process (Zimmer, 2018). In the heretical literature, where metabolic networks jostle to supplant genetic sequences (Capra, 2002:18), even the deterministic genome itself is said to be replaced as the fundament abstractive unit of heredity by the epigenetically tensioned cell (Noble, 2008 [b]). Orthogenesis, unpopular ever since

Lamarck's disciples ceded the battle for minds to Darwinism, is again becoming increasingly respectable - if suitably veiled by a diversionary vocabulary.

Establishment post-WWII genetics mandated that proposed drivers of heritable change fell within a rigid parameter of replicative error, a stance encouraged by and in turn itself reinforcing neo-Darwinian random mutation. Eventually, the advent of a discipline of natural ecology with its examination of trans-species communities introduced into biological theory dynamic networks of mutual dependences and active interrelationships, these insights propelling any number of explorative ideas more akin to prevailing Continental and Russian sympathies. Of course, whilst ever innocent of the nature of life, comprehensive elucidation of its transmission across generations must continue to elude comprehension. However, it is becoming increasingly plausible that a lifeform's epigenic encounters with vicissitude may, if not exactly dictate structural change, then at least intervene to realign, to release, reprioritise or repress, that is to edit, those (still metaphorical) molecular regulatory networks which would otherwise forbid further peripheral or elaborative expression.

2.4 Intelligence

Without exception, living beings behave intelligently, that is, self-determinatively within their peculiar ambit (Corning, 2014:242). This categorical and inclusive statement defies the fact that in the literature 'living' and 'intelligence' are alike cast as problematic properties made patent only by prescriptive definition derived from anthropocentric intuition and attribution; just as life is whatever criteria we care to apportion it (Lotka, 1956), so too is intelligence (Griffin, 1976). It is definition alone that unfavourably distinguishes innate behavioural response from a preferred educated intelligence exercised by experience and training (Schiff & Lewontin, 1986:8).

In this present context, and in so far as a self-aware consciousness (cognition) universally mediates vital teleological and self-preserving reactions to sensory irritation (Minot, 1902), consciousness and intelligence are inseparable, therefore effectively identical (Capra, 2015:245). Accordingly, because all living things behave intelligently and because it is impossible to define either animation or intelligence in such a way as can comfortably severally isolate but jointly encompass the entire spectrum of their expression, it is prudent to adopt the position that descriptors of living and of intelligence should be mutually transposable and so allow any concept of the one to imply both. To be quite clear:

> Living systems are cognitive systems, and living is a process of cognition. This statement is valid for all organisms, with or without a nervous system (Maturana & Varela, 1980:13).

Intelligence (or cognition if preferred) is therefore a cardinal property of life (Pollan, 2013), a causal factor in evolution and not something determined by it (Piaget, 1979:x). Some, nevertheless, still choose to intuit intelligence not as an inaugural component but as a subsequently evolved ancillary spanning a continuum from stimulus-compelled automatic reactivity through to a 'good enough' composite judgement, or perhaps still further to the conscious resolute responsiveness of rational calculation (Kahneman, 2012), each embodiment variously within a quintessential 'need to know' or likely 'able to sense' spectrum. In this anthropocentric *scalae naturae*, those behavioural manifestations of a living intelligence which most closely resemble mechanical reaction are separated by biologists into discrete, indeed contrasting, categorisations of tropic and nastic movement (in plants) or instinctive response (in animals), each an apparently unmediated reaction triggered by commonplace recurrent challenge and so deemed indicative of autonomic not intelligent behaviour. This presumption of innate mechanism often lies not in the behaviour itself but in the observer's description of it (Griffin, 1976), foreseeability of response rated as maladaptive when more felicitous bespoke options appear available.

Every organized being is self conscious (Trewavas & Baluska, 2011:1221) but the majority of biologists have long been reluctant to concede intelligent behaviours to lifeforms other than 'higher' vertebrates and especially so to members of the plant world (Trewavas, 2014). As instance of this prejudice, Cavalier-Smith (2006:998) contributes, 'The anus was a prerequisite for intelligence; without it, heads and brains would not have evolved.' No, not an anecdote promoting the English Public School, but a statement that betrays the unwarranted presumption that morphological complexity alone makes for a coherently intelligent life, lack of cephalisation denoting absent intelligence in all less sophisticated organisms including the entire plant and cellular world. Yet, as McClintock discovered, plant cells possess decision-making cognitive capacity to sense and react to unusual or unexpected events (Shapiro, 2010:2). Each living system is consciously aware of an outside world and of its own relative positioning in that perceptual world.

The preceding observation that all living beings behave intelligently, that is, responsively and appropriately within their particular ambit points to volitive behaviour, that is, to the organism as purposeful actor exercising degrees of freedom (choice) and not as mere observer of its own affairs. While it might be comforting for some to embrace estimation of purposiveness as a purely human attribute, it is abundantly clear that every level of animate expression exhibits both choice and anticipation, each a corollary of purpose. Volition denotes an individual and active, a moderated and particular interrogation of available situational options, the which, by subjective inference of relevance, attribution and prioritisation, enable an organism to actively advance purpose or to resile from harm, either disposition positively contributing to its own preservation and possible proliferation. Importantly, intelligence is elemental to life itself and not simply a facility only afforded to more sophisticated lifeforms. Consider, if intelligent action is inference from effect:

> An intelligible communication via radio signal from some distant galaxy would be widely hailed as evidence of an intelligent source. Why then doesn't the message sequence on the DNA molecule also constitute *prima facie* evidence for an intelligent source? After all, DNA information is not just *analogous* to a message sequence such as Morse code, it *is* such a message sequence (Thaxton & Olsen, 1984:211 *italics in original*).

Memory is an embellishment of this intelligent self, a resident prompt to revive or reinforce pertinent behavioural performance by dint of a previously schooled experience. As an extension of selfhood, memory thereby reaches beyond a domestic quarantine to project the organism into a more comprehensive reality that admits past, place and, if not a future, then at least the conviction of a continuing present. For example, by their symbolic homecoming 'dance', bees demonstrate memory of location, recollection of abundance, and expectation of continuity.

2.5   *Organism*

The organism is that minimal network of components necessary and sufficient for survival as a particular lifeform (Wake *et al.* 1983:221) and which, if multicellular, acts as a second order autopoietic system (Maturana 2011) within a rich repertoire of other interactive systems (Varela *et al.*, 1993:196). This self-referential individual, as functional representative of a species, is to human sensibility the discernable operational living unit, regardless of the several discrete contributing components that make such positioning possible. In terms of biotic organisation, the self-contained individual cell is a first order unit of living entity.[1] Obedient to this expectation, a prescribed threshold may therefore be set for a living system at the earliest point when a functional 'protoplasm', sequestered within an envelope insulating but not isolating its internal sovereignty, recognises resistive containment by a non-self (Fichte, 1794). For many biologists this autonomous undifferentiated cell is the quintessential organism (Haukioja, 1982:363).

Autonomy does not imply independence but is always relative or conditional. Whilst it may be superficially declared that the 'fundamental feature that characterizes living systems [*organisms*] is autonomy' (Maturana, 1975:313 *clarification added*), strict autonomy is just another facilitative fiction that robustly obscures the

---

[1] Providing living forms be conventionally if tautologically defined as cellular; *i.e.* thereby excluding viruses as being alive.

putative reality that every living thing, including the primordial eukaryotic cell itself, is either an incorporate composite (Wallin, 1927; Margulis, 1970) or obligate companion (Lederberg, 1952; Forterre, 2005) of another living form. The segregate life is a singularly communal strategy. Essentially, every unit of life interacts at multiple levels with others (Oyama, 2010:405), no one lifeform (nor tissue) can ever be fundamentally independent (Bohm, 1980:11). However, and despite elastic liaisons or the potential for conflation of autonomy with independence in guise of individuality (Huxley, 1852), a compromised ontology set by the operationally closed but purposive and autopoietic self retains signal status (Montalenti, 1974).

To perpetuate this sentient deployment, the self must be configured as a cohesive but interactive economy having persistent facility to reiteratively prosper by ceding its ancestral *baupläne* to ensuing and surviving generations: to parody Pross (2012:181) to perform as 'a replicating entity in a bag'.[1] It would, however, be a fundamental error and a fundamentally inappropriate scientific perspective to believe that the organism is

---

[1] While reproduction is not a necessary condition of a particular living organism, it is an essential prerequisite and necessary consequence for the survival of a living system (Maturana, 1974: 314).

merely an enduring DNA blueprint supported by a disposable body. Transmission between generations is by means of whole cells and not simply by chromosomes wrapped in a corporeal survival package; replicatory DNA, however it be defined or reasonably envisaged, is made the servant not the master of the living cell (Noble, 2015:11). In an intricate play of priority and influence, even the regulatory genes regulate by first themselves being regulated (Oyama, 2000:38).

The nature of the organism links into the long argued conundrum as to what may be the evolutionary unit of change - be it gene, individual, community, species or whatever. Simply put, upon what unit or subunit of life does evolutionary selection exert effect. The various solutions offered are, of course, as misguided as the question. Absent any legitimacy of natural selection other than the platitude that what cannot survive does not survive, it is the experiential operational unit (ostensibly a discrete lifeform) within its milieu as a bundle of conserved propensities that thrives or succumbs, reproduces or retires, initiates elaboration or encounters extinction. This said, however, it is ventured that when considering long-term evolution in an interactive elaborative regime it is 'almost impossible' to isolate one particular stressor from others because different types of influence and the several processes underlying their origin and transmission chronically interact (Jablonka & Lamb, 2008:393) making it insecure to identify any one exclusive, dominant or absent factor.[1] According to Lima-de-Faria (1988:268) the organism is a symbiotic evolution of the near-autonomous evolution of its components in separate but contributing processes - which is why fit, as a matrix of contributing evolutions, is never perfect but good enough.

2.6 Insinuation

Life has a particular ability to explore its own limits (Ruiz-Mirazo, 2004:325), each concretely autonomous expression, the organism, exerting both passive and active impact upon its own and others' milieu. Consequently, and whilst not immune to external compulsion, life in the act of living is never merely a toy of circumstance but is percontatorial, that is, positively inquisitive and actively intrusive. Far from being the beneficiary of a coincident survival, life exhibits the habit of an opportunistic exploitive

---

[1] For example, the dinosaur-eliminating meteorite.

expansion (Corning, 2014:247). For this property to be deployed, the organism's phenotypic embodiments need be accessible and sensitive to changed dynamics within its environ as well as pliable enough to adventitiously defy, to accommodate, or to exploit them. To these ends the quintessential expression of animate responsiveness, that is its plasticity and agency, is embodied in behaviour.

'Nothing is more obvious in the study of nature than the existence of complexity and levels of organization' (Hull, 1974:131). This exquisite intricacy is the captured effect of expansive elaboration canalised to lie within a *bricolage* of capriciously interactive associations, the outcome of one eccentric end proffered as further inviting means. An organism's reproductive memory, when coupled to its persistent capacity to sympathetically reposition to accommodate prompt of independent variables, permits novel exploitation of newly relevant competencies. Whilst such novelty at the gross level of the Type has for long been foreclosed (Kranich, 1999:2), inquisitive behavioural epigenic exploration is a continuing impulse actively and individually exercised by organisms in performance of their ordinary functional activities.

Epigenesis is that externally relevant, internally facilitated, and developmentally evinced variation, orchestrated by intimate experience and transmitted to subsequent generations, wherein responsive behaviours provide a necessary performative interface between the lifeform and its attendant stressful or liberating reality. Within such a reformist reality both enclosing environ and internal propensity have effect, a change to one impinging upon the other to promote fresh structural congruence. This theoretical perspective echoes the insight of Lamarck concerning the inheritance of acquired characteristics and takes place both outside of the DNA sequence and as a developmentally induced DNA variation:

> Contrary to the established view, soft inheritance is common. Variations acquired during an individual's lifetime can be passed on through epigenetic behavioural and symbolic inheritance. They can affect the rate and direction of evolution by introducing additional foci for selection, by revealing cryptic genetic variation, and by enhancing the generation of local genetic variations. Moreover, under conditions of stress, epigenetic control mechanisms affect genomic re-patterning, which can lead to saltational changes (Jablonka & Lamb, 2008: 394).

Indeed, immediate confrontation, enduring stress, or internal disruption, may well provide epigenic pretext but not merely as foci for selection as Jablonka & Lamb at first suggest but rather, as they immediately reaffirm, as an election to some ameliorative or audacious initiative.

Much foolish finger pointing has been directed at Bergson's imagery of an *élan vital* (1998). Even (and perhaps especially) within the Big Bang theory of universal establishment, it is likely that in both the inorganic and organic cycles of evolutionary development there resided a primordial proclivity to invest in differently elaborative expressions until such time as limit of prior path or bar of intrinsic property intervened to suspend or forbid further permutation. During the organic phase of this multiple evolution any subsequent exposure to a discrete but pertinent agitation might then have had effect to deliver the gratuitous relaxation of a particular constraint and so to enable a yet wider interactive novelty. It was to conceptualise in language this primal, potent and fundamental propensity for an active and expansive initiative rather than a negative and eliminative Darwinian piecemeal selection that Bergson coined the spirited metaphor of an irruptive *élan*, it however having unfortunate entail to conjure in less catholic minds a troublesome analogy to vitalism.[1] However:

> It is nonetheless a fact that, in general and as a consequence of the signification it acquired in the eighteenth century, the term *vitalism* is appropriate for any biology careful to maintain its independence from the annexationist ambitions of the sciences of matter' (Canquilhem, 2008:60).

Even Weismann felt obliged to subsume hereditable 'vital units' within a secular vocabulary as 'biophores' (1893:450).

Whilst accommodative behaviour circumvents individual elimination, an insinuating exploration recruits intrinsic biogenic ductility to foster intergenerational transfer within a reproductive milieu that favours resilience and rewards self-promotion. Absent a responsive behavioural competency, these are not dynamics available to material or to mechanical phenomena.

---

[1] The concept of a vital process as distinct from mechanical or chemical change has come to stay if still not yet sufficiently supported (Collingwood, 1960:136).

Part Three: Reality

3.1 *Perspective*

Reality is relative, is vested in the living and in they alone. Accordingly, and in so far as it may be perceived experientially only by mortal individuals, reality is at once transient, subjective[1] and contextual.[2] It is also largely taxon-specific or, in social groups, effectively consensual (Berger & Luckmann, 1967). In establishing reality no species may escape its own constitution nor fully comprehend another's and, because everything that is observed is sensed by a particular lifeform, there can never be a universally accessible absolute reality. Instance, for example, humanity's eccentric 'positionality' of employing our own species-specific norms as if they had universal

---

[1] Subjective knowledge as any substratum beneath rude appearance.

[2] Kant's things made manifest.

validity (Grene, 1995); indeed, was it not the essence of 'Einstein's revolution ... that we do not see things as they are but as we are' (Postman, 1996:20).

As a natural knowledge (Compton, 1984:356) rather than independent fact (Jones, 1982:ix) and as may be satiated by inquiry or contradicted by circumstance, organisms contrive their own reality. Consequently, unhindered by mind-independent standards, reality in a biological context is the embodied vantage[1] brought about by an organism's enacted co-determined history with its environment[2] (Varela *et al.*, 1993:2020) as further conditioned by individual experience and memory. Selfhood determines the reality of such an environment so that each lifeform and its media move in congruence until some irreconcilable event, condemning one, destroys both. No matter how close, then, to some absolute pregiven reality our own native insight may be, it becomes substantially less so when we pretend to 'stand in the shoes of others'. Thinking like a bat (Nagel, 1974) is, well, batty. Of course, bats live in a real world too; but it's not the same as our own.

---

[1] Normative reference.

[2] Increasingly, in place of a vague *environment*, 'the notion of *milieu* is becoming a universal and obligatory mode of apprehending the experience and existence of living beings' (Canquilhem, 2008:98).

In their upside-down reality, to parody Protagoras, Bat is the measure of all things.

Perspective is also a function of size and, perhaps particularly, of longevity. The miniscule organism whose life cycle extends to no further than the microcosm of a leaf in a single day, encounters, and can therefore only know, can react to, or be affected by, a very particulate dimension of reality. Natural processes and objects that exceed the narrow span of a species' sphere of attention avoid recognition,[1] and, just as awareness of a second law of thermodynamics has been reasoned to be unnecessary for a sparsely sentient or transient microbe (Sullivan, 1928), so a relatively super-intelligent organism whose life enjoys a longer time-rhythm than our own might find the theory not so much unnecessary as untrue. Recall, that 'Newtonian laws of motion hold good [*only*] for all motions whose velocity is such as to bring them within the range of ordinary human experience' (Collingwood 1960:24-26 *clarification added*).

History informs that it was the Classical philosophers who invited, if not invented, modern science by

---

[1] Hence the value of tradition and particularly of an inscribed history.

introduction of the concept of a manifest suprasensible truth (Bergson, 1998:347).[1] Yet, for all their famed sophistry, the Ancient Greeks did not themselves grasp that they were living in a dim and distant past. This antique error highlights both the privacy and immediacy of animate perspective, the which, like reality itself, is rooted in the here and now (or as a history,[2] the there and then reimagined in the light of the here and now). 'Time and reality are irreducibly linked' (Prigogine, 1997:187). Every living organism in each and every generation anchors itself to a home-spun telic reality in which the robust act of living and the experience of a lived-in world is personal, uncertain, and now.

Perspective, then, as an affective position isolated from within a cacophony of potential realities (Mannheim, 2015:174) not only imprints a delimiting mediation upon every observation, but necessarily interrogates the

---

[1] It is not immediately clear that inferred scientific fact can today be regarded as synonymous with suprasensible truth rather than merely suprasensory observation; it has, however, been claimed that it may have been the incrementally landlocked citizens of the Anatolian port of Miletus who effectively 'invented' science by their recognition of a natural agency acting independently of Divine mischief within the fast-silting and enclosing Meander Delta.

[2] And in paleontology.

integrity of human claim to an independent rational[1] science in that, whether as observer, participant, or as agent of inference, our calculation of the natural world is unavoidably and necessarily anthropocentric and functionally anthropomorphic (Guthrie, 1993). Or, more simply, culturally defined so that 'our ways of thinking derive from habits of organising our internal and external worlds as well as from particularly powerful assumptions, images, and traditions of enquiry and interpretation, confusions, history and so on' (Oyama 2000:123). We do not enjoy more reality than other lifeforms, our lives merely canvass different values (Canquilhem, 2008:119) and our every attempt to stand aside in deference to an abstract objectivity is itself but demonstration of a quintessential human ambition (Johnson, 1987): echoing Protagoras, Man really is the measure of all things (Jonas, 1966:23; Grene, 1974:200).[2]

Differentiation of the non-self from an otherwise pervasive all-self refines generic perspective (shared

---

[1] Rational science as synonymous with both theoretical and inductive science: the fundamental tenet of rationalism being that thought is an independent source of knowledge and a more trustworthy source than experience (Moser, 1987:28).

[2] Similarly, according to Nietzsche (2009), in the vegetable world, Plant is the measure of all things.

commonplaces) into an individual experiential reality that conscripts from an externality just those phenomena and events (stimuli) as make the performance of a particular *baupläne* possible and to which it will be sensitive (Barbaras, 2002:128, after Varela). It is within this subjective universe of ascertained certainty, or *Umwelt* (Uexküll, 2010), that an organism's sensory and evaluative competencies combine with practice and memory to circumscribe perspective.[1] This self-other dichotomy is a genetically determined relational perspective common to all animate realities (Laughlin & d'Aquili, 1974:115), but it was surely an extraordinary intensification (or naïve extension) of self awareness when the specificity of a metabolically reclusive autonomy resolved to communicate or cooperate with, rather than simply react to or capture, non-selves.

---

[1] Uexküll declined to propound a plant world-view (Marder, 2012:2) on the novel basis that animals being mobile possessed receptor and effector organs integrated by a nervous system whereas plants did not. A view compatible with the then conventional Plant/Animal Kingdom divide (Anderson *et al.*, 1984:11).

3.2   Perception

While perspective lends conceptual bias, perception is a real phenomenon (Dethier, 1964:1138) whereby, at the boundary between a lifeform and its externality, the organism samples, measures, or monitors its immediate placement within the wider scale of a global environment (Gibson, 1979). Behind this structural frontier, sampling and sentience conjoin in a manner too closely associated to be usefully separated since sensory detection without any facility to respond invalidates both.   Naturally, in any sensory experience, perception means intuit of something but not of everything.   Organismic perception and the object perceived are never isomorphic.   Importantly, sensory perception reinforces the asymmetric self-other divide, it acting as a filtered inlet for the outside world to confront an individual's precautious sovereign presence (Kranich, 1999:56).

Each animate kind has a particular array of receptive organs and interpretive models that vary between species (indeed, idiosyncratically between individuals) in terms of fidelity, acuity, reach, emphasis and functional relevance, the which in total attribute discrete but particular relational properties to what would otherwise be common phenomena. However, what specialised sensory modalities isolate to refine, internal cohesive process must enjoin to maintain the organism's coupling with its obligate habitat: utile perception therefore mandates possession of a discriminative editing, positioning, and prioritising constitution (Merleau-Ponty, 1963), an entailed functional intelligence[1] reactively engaging with effective stimuli to ascertain a still filtered, but significant and immediately relevant, reality.

Structural sensory exclusion[2] and a consequent inline edit of redundant stimuli, decisively combine to confine the conforming disposition within a chronic unreality that forecloses all possibility of any living organism obtaining a complete recognition of the total (or in human terms, a truly universal[3]) reality (Laughlin & d'Aquili, 1974:2). This embedded naïve species-specific subjective-objectivity informs and integrates sampled perceptual information into a part real and part contrived, a part fixed and part fluid *Umwelt* envelope,

---

[1] Bergson's 'Practical Intelligence'.

[2] For example taste receptors in the feet or photoreceptors distributed in the skin - let alone plant sensibilities.

[3] Any interpretation that is completely impersonal, explicit and permanent (Grene, 1974:17).

that is, into an operational universe as real as it needs to be by innocent entrapment in a workable reality which is at once inhered, bequeathed and constructed.[1] It is this real enough reality within a boundary of the organism's own making (Capra, 2015:244) and not some overarching impersonal global environment[2] that must be perturbed by any evolution-provoking disturbance.

Inevitably, cocooned by embodied sensibilities each within their own certainty, the majority of lifeforms do not in practice encounter (sense) novel situations, at least, not ones they have any hope to successfully confront. As a consequence, many thus reliably sequestered organisms avoid the tax of interpretive equivocation by needing only to detect invariant stimuli or, and especially in plants, to minimally distinguish relevant properties rather than to contrive a global representation of objects (Bergson, 1911:189).

Perception is of no value unless it presents discriminative and purposive opportunity to avoid, alter

---

[1] In a structurally determined capacity, found context, and relationally engineered milieu.

[2] Such the coupling, that without organisms there may still be a physical world but not an environment (Oyama, 2000:xiv).

or to accept the sensed scenario (Ackoff & Emery, 1972:100), for without a discerning motivation any reactive behaviour would be effectively chaotic and thereby self-destructive. Nevertheless, it is axiomatic that the organism can respond only to stimuli for which it is structurally and functionally prepared (Kranich, 1999:26). Accordingly, the criterion of sensory efficiency is that it be relevant to a lifeform's dynamic present. Relevance implies assembly of predicative patterns of reality in which objects of perceiving and action are one and the same, 'Organisms do not stand on or fly among images and representations' (Turvey *et al.* 1981:239) and patterns do not exist until they are realised by reciprocal selection or coaction (Oyama, 2000:35) in which perception, preconception and memory are intimately and inseparably enjoined (Jacob, 1982) to such extent that an intended event flows in time from a satisfied anticipation (Kelly, 1963). Everything beyond the perceptual is conceptual; percepts without concepts are blind[1] (Portman, 1990:37).

Whatever configuration pertains in a wider environment, awareness of it is filtered by sensory edit such that animate reality is redefined by each organism's

---
[1] Adopted from Kant (1787:75).

own cognition of it (Ho 1991:348). In this regard, the principal contribution of science to human progress has been the utility of accumulating data which are found outside of ordinary human experience (Minot, 1902:4). This extraordinary extension of sensory experience (and therefore of the human *Umwelt*) has had substantive effect not only as accrual of knowledge but, and more significantly, has led to the development of technics enlarging the limits of inquisitive exploration and so of resource appropriation (Moore, 2015:151).

3.3 Meaning

Meaning adds functionality to perception by providing a protensive narrative (Husserl, 1991) for those behaviours that position the organism within the substance of its own milieu. As an implicit species-specific envelope within which reality unfolds, meaning is obtained not by naïve internal reconstruction or from reflection of an already sharply edited external world but likely by sift through resident presuppositional encoded templates[1] (intuited resemblances), ordering associations by regularity and coherence into functional

---

[1] Coded in a sense akin to Deleuze & Guattari's (1993) notion of *agencement*.

relationships with effect to pattern (categorise or reify) and thereby to discover salient anticipated or anomalous but actionable, that is meaningful, data. For example, to register contiguous antecedent cause (agent) from presented effect (consequence).

Associative meaning derived from inhered template may be further extended as metaphor across domains (Jones, 1982:51) to verify or to apportion acquired experience. Metaphor intrudes somewhere between insight and illustration when either warrants distributive transfer between concepts. As such, it is not merely a tool of language but an element of systematic mapping available to all dimensions of animate experience (Lakoff, 2000) and of which human practical and propositional language merely takes advantage. Of course, the linguistic infelicities of metaphor can not figure in the meaning-making inferences of, say, an insect, but neither does it forbid essentially the same bridging of domains to exert an equivalent generalising effect to realign or to stabilise the realities of organism and species (Lakoff & Johnson, 1980). Metaphor changes reality by changing the observer's sense of it, adding associative nuance to already calculated evaluations.

Just as chemical elements naturally fall into predetermined modes of affinity and reaction, so do animate categorisations of stimuli acquiesce to embedded fundamental and simple rules for making rules, the application of which not only fashions perception into meaning (Avital & Jablonka, 2000:74) but also attaches a gloss of clarity that allows coherent associative memories to condition experience (Schlicht, 1998). Whether this gloss of meaningful conviction discovers or imposes order is less clear, but in either event the effect is to conjure a manageable subjective-objectivity that is not only species and capacity relevant but, especially, is plausibly (good enough) reliable.

Without this ordering narrative, sensory 'data' is merely noise. Accordingly, because raw data is superficial in its disclosure of the nature of things (Whitehead, 1968:7) and to the extent that a lifeform's own sensory receptors may not by their very structure or inhered receptivity have already substantially refined the stimuli (Merleau-Ponty,1963:13), organisms intuit meaning by ideation of consequence. Emotive sense for consequence derives from an innate[1] appreciation of linear cause, for without confidence in a causative

---

[1] Innate in the sense of an unlearned but functionally reinforced nativism (Mameli & Bateson, 2011).

sequence of events no reactive behaviour could ever be worthwhile. Prescribed or anticipated consequences combine to prioritise attention and direct action. Commensurate with its embodied coupling and from this predicate of a prevailing dynamic the organism must orchestrate, from awareness of its internal state and a sensed external reality, all such complementary responses as might prove predictably appropriate for purpose. Operational transcendence of both raw stimuli and intimate reaction occurs primarily within that noumenal space attributed to consciousness; quintessentially, animate super-sensory consciousness is a meaning-making prerequisite having effect to mediate (manage) the complex logistics of a coherent goal-directed motility within the lifeform's own reality (Merker, 2005).

Living things take life seriously; they 'are actors, not merely patients, of their own life history' (Mpodoziz, cited Camus, 2000:217).[1] Everything intruding into a subjective *Umwelt* is either afforded a meaning-quality or is otherwise peripherally, but effectively totally, neglected (Uexküll, 1982:30) so that in this manner nothing of consequence is incongruent with a lifeform's world view. All else effectively invisible or irrelevant.

---
[1] Actors as opposed to *reactors* per West-Eberhard (2003).

For each organism, then, perspective, perception, and practised participation combine to immerse it in a reliable and actionable, that is a meaningful, reality. Undoubtedly, the world extends beyond the reach of mind both in fact and in imagination (Fodor, 1983:120; Nagel, 1986:90) such that neither our own nor any other animate sensibility can do anything but pretend to apprehend the infinite potential of a truly objective perspective.

The mindful human animal lives in its own orbital *Umwelt* of practical necessity and perceptual subjectivity but, and because of particular cortical and languaging capacities, is able by conversation and technology to expand, if only indirectly, perception to the imperceptible and so to greatly embroider experience. In particular, acute ideational skills lift the species onto an experiential plane from which only cogent rationalisation can obtain unproblematic agreement between exigency and expectation (Canquilhem, 2008:xix). Indeed, that humans are more rationalising than rational has, uniquely, led their native animal curiosity to investigate levels of meaning which could only be accessed by adventurous speculation, severing the direct link to a lived reality with potent effect to accentuate invention. However, beyond need for an eliminative abstraction in calculation, the scientific materialist fiction of a rational, objective, permanent and unconditional deterministic certitude remains quite invalid[1] (Toulmin, 1992:21); for, as Wirth (in Mannheim, 2015:xxvii) advises, there is no value apart from interest and no objectivity apart from agreement; these, metrics of meaning-making common to all animate realities.

3.4 Anticipation

Anticipation invests memory with utility by elicitation of educated response to a recurrent presentation (Avital & Jablonka, 2000:70) and, as premiss for purposeful behaviour, supports the normative expectation of a natural continuity between past, present, and future (Trewavas & Baluska, 2011:1222). A reality without confident anticipation of temporal reliability would provide no sensible motive for any action the prospective control of which must be envisioned ahead of time (Hofsten, 2014).

Neither innocent expectation nor deliberated prediction is useful unless either unfolds within the familiar and

---

[1] Or, for that matter, the life everlasting.

manageable parameters contained by a particular organism's accommodative powers and captive reality. Accordingly, feasible anticipative physical action is always and only towards those parts or whole objects (animate or material) perceived to comprise or protrude into the composed space of the organism's *Umwelt*, this discretionary arrogation made without regard to any wider recognition that each such recruited object may have thousands of redundant qualities of no concern to the subjective observer (Uexküll, 1982:72). Those peculiar properties discerned to attune with an observer's dispositional utilities are referred to as affordances (Gibson, 1979), each the operational complementarity of an object's characteristic to a particular purpose of the observer. Taking advantage of this complementarity goes beyond mere anticipation, it signals intention.

Even the apparently most primitive of motile organisms exercise purpose (Nakagaki, 2001), integrating the direction of inner motivation with such affordances as a presented situation permits (Marijuán *et al.*, 2010) to instigate informed anticipatory behaviour within a constituted but reliable reality.

3.5 Communication

In an otherwise unlabelled world, communication, whether it be expressed in ritual, as representation or record, by touch, sight, sound, gesture, scent or symbol, at once projects and supplements individual understanding. Importantly, each act of intra- and inter-species signalling predicates expectation of realities held in common. Accordingly, all meaning-carrying communication signifies biological (operational) interdependence, although a distinction might usefully be made between those signals that independently stem from superficial or incidental structure (say, colour or conformation) and those which emanate from active behaviours (say, vocalisation and performance).

Meaningful communication reaches beyond mere transfer of information (Polizzotti, 2018:7) to have intrusive effect upon the *Umwelten* of all involved and therefore to establish an immediate congruence with both the communicator's own reality and those of a wider population. This dynamic sharing of realities is finessed by purpose to either persuade or deceive but usually to inform by, crucially, the communicator recognising appropriate powers of comprehension within an anticipated audience of 'others'. As a volitive behaviour, communication conscripts attention and intensifies interaction across personal and species barriers, a facility substantially pronounced, if not essentially different, in the human animal.

It was long claimed that because anchored by their sessile[1] nature, plants rely on structural plasticity rather than active communication to register recognition of an exigent reality and so invest nothing but a solitary response to those pertinent influences made manifest within their own isolate vegetary environ (Silverton & Gordon, 1989). It is an error, however, to confuse motion with locomotion; plants continuously explore discrete but complementary rooted and aerial relationships without the need to scurry about to do so (Marder, 2012:1367). Fortunately, plant communication is no longer a taboo concept (Trewavas, 2014). It is now securely established that plants variously sense and optimally and positively respond to kin and competition, to variables of light, atmosphere, water, gravity, soil structure, vibration (sound), nutrients, toxins, microbes, pollinators, herbivores and physical obstacles as well as to chemical signals emanating from other plants and root mycorrhizae.

---

[1] Sessile does not mean stupid (Marder, 2012:1365).

That these alarums and conversations occur on an extended timescale does not invalidate either their nature or efficacy; nor does it preclude elective awareness of an *Umwelt* with its entailed connotations of anticipation, attribution and, in particular, appropriate and often broadcast purposive activity. Evidence is fast accumulating that on the single cellular organismic scale, too, signalling between individual bacteria and social neighbours occurs as a regular and essential function, proportionately 'fit for purpose' intelligent intervention evidently occurring at all 'levels' of animate existence (Jacob *et al.*, 2004).

As a volitive behaviour, purposeful communication adds persistent nuance to the positioning of disparate lifeforms within a motley of interlocked *Umwelten* realities, their chorus of extended and combined sensibilities having potential to reflect or deflect interactive epigenic expression.

Part Four: Behaviour

4.1 Interface

The world is, and it is the function of an operational intelligence to find within it such manageable order and sufficient reason as best allows a profitable navigation. Behaviour is the autonomous disport of an organism aiming to produce results in that outside world (Piaget, 1979:144), its performance primarily relying upon a concept of causation (Michotte, 1963:183) and its signal effect to commit a species towards efficacy of particular habits. Pivotally, behaviour is the interface between life and living.

Whilst life is sustained by an immediate and internally focussed discipline, living is a forward and externally orientated dynamic. As primary actor in the act of living, behaviour variously initiates, accommodates, or pre-empts disturbance by intelligent exercise of a structured plasticity (that is, accessing organisational pliability by expressing it) as well as repurposing elements of its environ to achieve specific ends (Godfrey-Smith, 2017). Essentially, behaviour operates to favourably position the organism within a perceived milieu (Compton, 1984:357) by both internalising external stimuli and by externalising those internal states which act to manage a surrounding world (Oyama, 2000:26), exercise of choice providing continuous mediation between the self and its *Umwelt*. However, if a lifeform is confronted by some inordinate or recurrent response-defeating tension, whether ultimately adverse or advantageous, both the lived experience of challenge and the utility of its own responses resonate throughout its entire being, educating intelligence, marshalling metabolism, informing reproduction (Nikolic, 2015:46) and disturbing organisation with effect to annotate future genomic expression (Piaget, 1979:142).

It is *via* this epigenic reverberation that confronting imperatives, internal systems and reproductive regulators are brought to mingle on intimate terms, impinging effect reaching beyond the block of behaviour to access constitution. This habituated practopoietic[1] intimacy ensures the realisation of a

---
[1] Animate capacity for self-adaptation to newly emerging contexts.

continuity in which individuals may be relatively expendable but lifelines resolutely less so. As both cause and consequence of evolution (Plotkin, 1988:8), behaviour has potential to change the relation, but not the relationship, of cell, its DNA, and its reality (the organism's structural coupling with its medium). In essence, therefore, it is by causative (that is, anticipative and purposeful) behaviour that individuals seek first to circumvent, then to accommodate or to exploit challenge. Chronic exposure or insufficient effect internalises generational (conspecific) experience and may prime parental reproduction to an extent that progeny can develop to better address it; consequent evolution of novel form a solution not the process. Process is living the percontatorial life.

4.2 Purposive

Paradoxically, whilst neo-Darwinian science relies upon the mathematics of a random genetic fortuity (Monod's chance and necessity) to indirectly determine direction, genetic activity is itself an implicitly purposeful process (Turner, 2009:206). By their responsive behaviour living things positively orientate themselves within their found situation, a deportment not just confined to animal species but a defining characteristic of all lifeforms. The incidence, function, speed and expression of this intent alignment necessarily depends upon dynamic context, on relevant affordances, the behaving species' inhered structure, as well as their irritability or capacity. Much behaviour is in reaction to external cue, still more in anticipation of achieving a private objective.

Strictly, of course, a salutary effect is not of itself sufficient evidence of purpose but neither needs teleonomy[1] be intellectual treachery:

> We present the concept of gravity only as a derived image, without speculation on its nature, and this is justified, for the introduction of such a term serves to simplify and unify our concepts. We insert it and analogous physical terms into the laws of causality, which clarify the appearance of certain changes for us; we do not use the terms to say anything about the essence of the forces in effect. *Likewise, in biological fields, we must take the basic phenomena of life into account and use them in our deductions*, even though for the time

---

[1] Interestingly, the Oxford English Dictionary (1989:728) defines teleonomy as, 'The property, *common to all living systems*, of being organized towards the attainment of ends' [*italics added*]. Monad (1971) proffered the term teleonomic to signify purpose without patent first cause.

being we cannot determine more precisely or explain the mechanics (Schindewolf, (1993:374 *italics in original*).

In a justifiable science, then, gravity is what it does. Life, too, is what it does. Importantly, a stance to comprehend life not by what it is but by what living things do goes far to assign to behaviour the pivotal role in evolutionary development and therefore, given that a presented behaviour may be particular to the individual, effective evolutionary impress must find expression within an already determined *Umwelt*. Accordingly, should a non-catastrophic external event prove irrelevant to the lifeform's contained world it exerts no influence for change. Of note, wholesale destruction is no more an evolutionary force than extinction is adaptive.

The cardinal phenomenon of life that we observe is its purposiveness, itself an incident of governance by sensory and motor control systems which makes responsive adjustment both to and within a milieu possible (Gelman, 2009:248). However, such reactive governance is directed by expectation of congenial outcome and because thus positively end-directed, behaviour exhibits as a quintessential purposeful agent (Corning, 2014:242). Together, purpose and prediction enjoin in behaviour to pre-empt a future that would otherwise be inordinately dangerous. Purposeful activity is not, however, confined to practical survival but extends to the inquisitive mastery of a found environ (Piaget, 1979:6), to deliberate inaction, as well as to playful entertainment and rehearsal (Shaw, 2002).

It has been a prejudicial Cartesian dowry to deny purposive action to all but human actors. Of late, however, extension of goal-orientation to 'higher' animals has gained wider approval and, but only with ample caveat, may now even be tentatively and selectively accorded to other (particularly mobile) complex and singularly un-human organisms (Godfrey-Smith, 2017:59).[1] Nevertheless it still remains prejudicial to professional standing to too enthusiastically attribute purposive behaviours in plants. However:

> If we continue to define behaviour ... as goal-directed action designed to use or transform the environment, or to modify the organism's situation vis-à-vis the environment, then it is clear that

---

[1] With regard to their patent sentience, Godfrey-Smith advises, 'octopuses have often been listed as a kind of "honorary vertebrate" in rules governing their treatment in experiments, especially in the European Union' (2017:59).

there is such a thing as as plant behaviour (Piaget, 1979:125).

Despite recent (and not so recent) advances, anthropocentric sentiment continues to restrict behavioural research by respecting the (biologically irrelevant) uniqueness of Man - as if frogs and fir trees were not just as singularly unique - and as purposefully active.

4.3   *Appropriate*

Appropriate behaviour is spontaneous (Ho, 1991:337) and evidence that the self-managed organism is also managing, as best it can, its own environment. The positioning function of behaviour implies an envelope of action within which an organism's generic structure permits inquisitive access to an appropriated domain of tolerant interaction between the found niche and its own sensory and effector surfaces.   In this conscripted[1] space largely engineered by the organism

---

[1] Conscripted by sensory reach, mobility, and imagination.

itself (Jones *et al.*, 1997),¹ apposite fit-for-purpose behavioural objective is best achieved when resolute act and facilitative affordance coincide in a performance wherein achievement matches intention, that is, when action orientated towards an objective may be efficaciously and reliably realised. Efficacy implies grasp of anticipated outcome. Affordances are those apperceived invariant properties of an object or situation which invite, suggest, or recall in the observer an appropriate, that is a learned or embodied, predicate recognition (Gibson, 1979): for example, trees do not suggest climbing to pigs in the same way they may to monkeys. Strictly, trees do not suggest anything, they are simply there, framed within the world view of a particular observer.

Critically, it is behaviour that mediates the dynamic between an autopoietic lifeform and impinging independent externalities, appropriate (that is, life-supportive) action the central motivation of relational interactions which reach into the very core of an individual's constitution to variously corroborate, stimulate, prompt, or to forbid epigenic effect. Without

---

[1] Strictly, 'it is difficult to imagine a life strategy that does not in some way lead to a degree of modification of the abiotic environment', nor of some intervention in the availability of resources to other species (Wright & Jones, 2006:203).

such sympathetic alignment, brash confrontation with encroaching external forces must prove too common and too abrasive for an unresponsive species to withstand. Equally, this same activity-driven constitutional flexibility admits novel opportunities which, again, would otherwise evade the unresponsive actor (Rensch, 1960). Behavioural studies do not lend themselves to controlled experiment, but it may reasonably be surmised that no matter how an observer seeks to construe it, an organism's conduct takes place within its own contrived *Umwelt*; appropriate action thereby circumscribed by available native competencies discriminately exercised within accommodated or accustomed parameters.

4.4 Experiential

Behaviour mediates between a lifeform's conformation and its operational circumstances with effect to best harvest advantage from these twin endowments as well as to circumscribe such sporadic risks or novelties as may prove peculiar to its provincial non-self world. Experiential learning, that is inquisitive behaviour coupled to memory of past encountered consequences, customises and usefully enlarges this native capacity by addition of a self-tutored prescience or practised dexterity to the task of living and, in subsequent generations perhaps, to ingrain fresh nuance into an already embodied repertoire of latent or associative potentials. As unambiguous evidence of animate participation in external affairs, educated encounters have effect to increase the number of an individual's interrelationships and so to both refine embodied perspective and to adjust the elastic envelope of its *Umwelt* (Piaget, 1979:136-7).

Essentially, a trained rat is a changed rat.[1] Learning is not simply a tutored progression from profound ignorance to smug certainty but, instead, acts to attune an organism's resident repertoire of possible behaviours to address recurring or persistent situations that casually re-emerge in a largely stable environment. However, this flexibility has its limits; as 'Lorenz and Tinbergen had been insisting for decades ... learning is constrained and not an all purpose generalist capacity' (Plotkin 1988:135). A flea cannot educate itself into being an elephant; organisms are primed to learn certain kinds of relationships and not others. However, it follows that if, for example, a primate can successfully learn and apply a workable lexicon of human sign language it is because

---
[1] A pertinent notion attributed to B. F. Skinner (Ho, 1991:349).

it already possesses a connate capacity for symbolic communication.[1]

Because experiential learning is an affective individual acquisition it mandates awareness of a selfhood which admits of its own instruction, and since experiential learning can be demonstrated across the spectrum of protoplasmic and metazoan organisms then self-awareness itself must also reasonably be deemed a universal animate attribute. Further, because memory does not retain raw sensory data but instead disposes relational meaning, self-related learning and subjective interpretation are inseparable. Experience, then, happens only to a self-referential interpreter; a status that necessarily, if differentially, includes all living forms.

4.5 *Obligate*

Obligate (that is, apparently unlearned) behaviours assigned as instinctive in animals and as tropic or nastic in plants are those impelled or reactive activities triggered by compelling or recurrent challenge

---

[1] Instance, orangutan Chantek of Zoo Atlanta.

premised solely by stimulus[1] and for which many human observers deny intelligent or rational mediation on grounds of the actor's structural or neural inferiority. Accordingly, inception of these observer-defined behaviours are deduced to be mediated by neither cross reference to secondary sensory inquiry nor by recruitment of conscious attention, but rather display as pre-formed and inevasible attributes of an organism's core organisation. Apparently, given the endemic variety and sophistication of such notionally unlearned behaviours there are instincts for everything, they together calling upon an encyclopaedic repertoire of archived responses many an endowed cognitive animal would find mentally cumbersome.

Much ostensibly obligate behaviour extends well beyond immediate reflex to prime extraordinarily precise reproductive or migratory performances as well as intricate chains of ordinary events, some of which may exhibit once in a lifetime or between generations, others as repetitive activities primed by association, reinforced by experience, or avoided by habituation.

---

[1] This nice notion of a sensory premiss is employed by Plutarch (circa 46 -120 AD) in *Moralia: de sollertia animalium*. Following anecdote of fox and dog behaviours he proposes, 'Perception here affords nothing but the minor premiss, while the force of reason gives the major premisses and adds the conclusion of the premisses' (1957:329).

Part Four: Behaviour 133

Always, of course, these obligate behaviours are expressed by autonomous organisms not automatons, and it remains extremely unlikely that science has yet even begun to properly conceptualise the true nature of this genic logic and the degree to which it underlies or determines mechanisms of otherwise apparently rational evaluative cogitation (Lefebvre, 1985).

Frequently, designation as instinctive, when regarded as an invariable and ineluctable species-specific characteristic, is no more than the observer's insecure inference that because a particular behaviour presents as routine it must therefore be innate and, if merely innate, then potentially dysfunctional.[1] To some degree every lifeform exhibits a repertoire of stereotypical behaviours within particular situational parameters, any further discretionary powers of selective attention, of memory and cognitive deliberation secured and deployed by tiered layers of processing or of substitution that were likely elaborated to expand, to weigh, or to manipulate perspective with effect to better address the intricacies of a more generously explored and therefore

---

[1] Categorisation of a behaviour as instinctive blocks further investigation by substituting labelling for explanation (Johnson in Oyama *et al.*, 2001:16).

increasingly surprise-prone physical and biosocial world (Kahneman, 2011).

Resident genic knowledge in overtly intelligent animals regularly presents in display and ritual performances, most usually but not only associated with mate attraction, parenting and territorial delineation. Imprinting, for example, is a learning-informed situation-triggered instinct. An instinctive proclivity for complex skill-driven behaviours has been observed in chimpanzees born and matured in single sex laboratory cage conditions and who, on release as adults in a mixed group into a wooded reserve, readily adopted tree climbing and other 'natural' behaviours.[1] Prairie dogs in zoos with hundreds of people streaming past still 'instinctively' post a lookout (Kropotkin, 1902).

At this point it must be emphasised that all attribution of animal mentality is necessarily if insecurely informed by our own introspective understanding of both the human mind and of human motivation - each a highly contested and irretrievably flawed exercise of subjective philosophical or psychological presumptive logic (Varela & Shear, 1999).

---

[1] As part of a USA Federal program to de-commission experimental chimpanzee models.

4.6 Communal

It has already been observed that the segregate life is a singularly communal strategy, every lifeform either an incorporate part or the obligate companion of another. This structural affinity extends in many species of organisms into wider advantageous temporary or permanent kin groupings or social communities of conspecifics that variously function to enhance individual efficacy, to concentrate effort, or to mutually avoid a plethora of hazards. There are likely many skeins of extended communities of which we are quite unaware - only relatively recently have fungal mycorrhizal soil networks begun to be understood, for example. Depending upon the observer's interpretation, conjugations of individuals might reflect either a community of interest or interests held in common; maybe herds simply go where the grass is and do not necessarily gather together for the personal consolation of company. Of course, once regularly assembled, a new elaborative momentum is made possible, complementary opportunities opened to exploration and solitary needs the more readily communally satisfied or aggravated.

The natural world is not obliged to conform to the categories that our concepts and vocabularies seek to impose. Instance how sophisticated colonies of ants and termites with their familial ties, anatomical distinctions, division of labour, and complex continuous communications challenge both our understanding of the autonomous self and that of a personal identity. In fact, such examples of social unity undoubtedly mask an extension of animate organisation and are not simply cases of distributed or cooperative individual endeavour, for, as Marais (2009) proposed, the colony is the organism, the members' actions its active limbs. Human societies are less rigidly contained, and then primarily only by a fluid culture, itself an amalgam of kin relationships, regional loyalties, and geographical boundaries, (prototypically) cemented by a shared belief or value system reinforced by social pressure or coercive policing.

This present study does not pretend to account for the interminable variety and intricacies of animate behaviours, but is rather the recitation of a thesis that evolution occurs when the percontatorial life licences novel modes of living to form fresh congruence with emergent states. Although no privileged cause logically pertains to the tumbling trajectory of this living multiway network of interactions (Noble, 2015:7), volitive behaviour nevertheless remains central to evolutionary process. By their behavioural responses, organisms can convert a novel contingency into a fresh reality the which, if it effectively dispels the architecture of prior *Umwelten*, constitutes an elaborative development.

Part Five: Recollection

The affective world consists of the properties of objects made accessible to us through our senses. Such perceived properties do not exist in isolation but manifest either the nature of material things or are imaginatively projected by us as rationally necessary to sustain a coherent worldview. However, the often irrational nature of this same explanatory rationality admits a level of dissonance sufficient to provoke a periodic insubordination that both science and religion separately vie to claim and contain. In the 19th Century a newly crystallised topic of evolution was recruited to bear the brunt of dogmatic differences - not simply those of science and religion but between rival notions of science. From these multilateral disputes Charles Darwin and his ecumenical theory of natural selection emerged as biology's appointed emissaries.

Formal exposition in science begins with a prejudice (or an enthusiasm) framed as a hypothesis (Wirth, in

Mannheim, 2015:xxii) most often to be predictability authenticated as proven or disproven by demonstration, by inference, or by argument. In hypothesising a notionally contestable theory of evolution, the two textbook progenitors, Lamarck and Darwin, speculatively adopted diametrically opposed but equally arbitrary mechanisms of progressive adaptation. On the Catholic Continent, life was accredited its own constructive agency; in Puritan Britain it was pressure of competitive survival in a harsh market of obdurate externalities which set the agenda for adaptive change. Their antique polarisation between intrinsic or extrinsic momentum needs not detain us. Both Lamarck and Darwin adventurously operated in conditions of total ignorance of any science of heritability, indeed, effectively of any biological science at all beyond that secured from rustic anecdote, traveller's tale, or the exotica of curiosities cabinets.

It has since been proposed (Lima-de-Faria, 1988) that there have been three sequential, autonomous but obligatory evolutions, those of the elementary particles, of chemical elements, and of minerals; each new ordering conditioned by the circumstances of prior canalisation and all realised prior to a subsequent, fourth, biological evolution, itself sufficiently constrained by antecedent effect as to narrowly turn upon elaborative self-refinement. For this last effect to have occurred in the manner that it did, a mature materiality acquired fresh impetus and direction by felicitous conscript of an independent episodic drive to extend passive repetitive cycle into active linear process, it leading to the novel metamorphosis of a self-referential operational functionality that rests upon the intelligent management of self and surrounds *via* behaviour.

Essentially, it is this resident intelligence that separates life from non-life, the percontatorial character of which transmutes structured form and function into the dynamic of a living organism. At the interface of life and living, conserved formation and operational behaviour position the organism within an inhered milieu. Thereafter, the individual efficacy of inquisitive and reactive behavioural responses determine opportunity or detriment, enabling each act of living to retain congruence with a shifting contemporaneous reality. However, and whenever opportunity and capacity coincide, life may admit a fresh mode of living to occupy some new space within an emergent relational context. The historical procession of these fluid former interactions is evidenced by fossil trace and envisioned by the sympathetic human observer as a natural (that is, a rational) evolution in contradiction to the (apparently only) alternative explanatory notion of an irrational if guided deistic creation. As agent of effect, and standing apart from the myriad of contributory secondary or occasional impacting influences, behaviour presents as the quintessential and principal active determinant of animate evolutionary process by, *inter alia*, enactively realising or enhancing structure. Behaviour prepares, positions, promotes, protects and perpetuates life in the infectious excitement of living.

Life lives off its own. Unlike energy-consuming fire which can be lit by several methods, in various media and at any time, energy-manipulating life 'can progress only by means of the living' or, tangentially, expresses only in self-generative acts of living.[1] Contrary to the rapid chemistry of ephemeral fire, the perennial act of living conjures its own fuel from the molecular mining of a mineral substrate enlivened by resident metabolic process into self-reproductive lifeforms the which, thereafter, are themselves ingested by other organisms in a self-sustaining or minimally dissipative cycle of destruction and renewal efficaciously churning at such

---

[1] The organism as life's phoenix (after Benson, 1989).

pace as may permit the productive dynamic to be elevated onto an interminable conveyor.[1]

Experiential and vicarious learning allow organisms to shape, that is to customise within their structured capacity, each their own *Umwelt*. Humans, as articulate animals, have enhanced capacity for a fertile imagination that may amuse or illustrate but need not always overtly inform. As perceptual animals we live in a reality that is co-established by us and which may only be extended when we engineer or are ourselves engineered into new physical, cultural or technological relationships. We may modify or mollify but can neither escape nor stand aside from this blinkered human reality. Importantly, once the conceptual Rubicon is crossed from creation to evolution, issues not only of deities and miracles but also of absolutes and objectivity become equally contentious; religion and science thereafter continue to coexist as parallel repositories of both certified and conjectural explanation.

---

[1] Akin to Schrödinger's 'negative entropy' cycle (1967:70).

For tool-making Man,[1] science has proven a cognitive implement of extraordinary utility. As an instrument of both exploration and of explanation, the methodology of science marshals curiosity by rationalising experience to expand expertise. As expressed by Minot, 'The principal contribution of science to human progress is the recognition of the value of accumulating data which are found outside of ordinary human experience' (1902:4). As technology, this inference of the invisible tends to reposition scientific inquiry towards the 'how' and away from 'why'. Inevitably however, as cultural constructs, both conceptual and applied science are taxed by those same insular irritations which ineluctably inhabit the politics of place and person, for while a parochial perspective can admit diversity of opinion it may also permit a provincial originality to be brusquely dismissed or bullied towards some dominant orthodoxy. Fortunately, fashions in human curiosity encourage the cyclical reinvention of a previously discarded or derided intuition to freshly interrogate the received paradigm. This has been so in evolution theory. Except for the fanatic few, and

---

[1] Tool-making is not a signal capacity of the human animal but has uniquely imposed upon it an irreversible dependence upon technology (Kingdom, 1993). The tools of linguistic symbolism and aesthetic perspective are products of brain function rather than of structure-facilitated manipulative dexterity (Mumford, 1965:924).

although blind for more than seven decades to the wisdom of Waddington, evolution biologists are no longer 'entirely satisfied that nothing more is involved than the sorting out of random mutations by the natural selective filter' (1942:563).

Life, when incorporated as or into a living organism, is responsive to those conditions necessary for its prospering. On encountering a sufficient and stable matrix of benefits, familial radiations collude to form a common habit of exploitative speciation. Later, should some persistent internal or external circumstance occur that lies beyond the accommodation of a species' honed congruence with its surrounds, developmental recourse may be had to that archival labyrinth of pathways which signal its particular constitution. Accordingly, periodic regression in order to progress is a necessary feature of biological continuity. Unlike behaviour, continuity does not recruit intention but instead, seizing metaphor from analogy, reflects the interaction of discrete variables compelled towards a common equilibrium in much the same way that a drop of ink positively invades a sheet of blotting paper until the absorbency of the latter negates the concentration of the former.

Given the complex foundation of every lifeline, structurally determined functional originality arises primarily from reconjugation or reprioritising of already conserved ancestral periodicities retained within a genome, these proffered, accessed or variously impelled by intuit of emergent relational opportunity or impediment. For example, it is established that common gene regulators are present (if variously applied) across a wide spectra of animate phyla; instance, an homologous gene engaged in the eye formation of vertebrates, ascidians, insects, and cephalopods, and so likely accessible throughout the entire range of the metazoa (Halder *et al.*, 1995:1788). As complicit partner of opportunistic elaboration, past genetic disposition or discretion may be adventitiously revisited or recombined in any such renascent form as was not irreversibly prohibited by some prior divergent configuration, an accreted organismic past solicited to re-engage previously unexplored or presently muted possibilities as might enable a challenged skein of life to survive, to reorient, or to further elaborate.

Familial speciation is the product of a group kinetic exploiting a local stasis, any subsequent novel revision or reversion to a previously unrealised or retired format and characteristic, the contingent effect of an emancipated or obstructed momentum. Both intuition of crisis and a consequent triggering of developmental change derive from the feedback of facilitative behaviours, that is, from epigenetic shock at the interface of life and living. In essence, behaviour cradles life within a managed reality.

References

Ackoff, R. & Emery, F., 1972, *On Purposeful Systems*, Aldine-Atherton, Chicago, IL.

Anderson, M. et al., 1984, 'A semiotic perspective on the sciences: Steps towards a new paradigm.' *Semiotica*, vol. 52, no. 1, iss. 2, pps 7-47.

Avital, E. & Jablonka, E., 2000, *Animal Traditions, Behavioural Inheritance in Evolution*, Cambridge University Press, Cambridge.

Baluska, F. et al., 2004, 'Eukaryote and their cell bodies: cell theory revised', *Annals of Botany*, vol. 94, pps 9-32.

Barbaras, R., 2002, 'Francisco Varela: A new idea of perception and life', *Phenomenology and the Cognitive Sciences*, vol. 1, pps 127-131.

Barry, G., 2013, 'Lamarckian evolution explains human brain evolution and psychiatric disorders', *Frontiers in Neuroscience*, vol. 7, no. 224.

Bateson, G., (1972), 'On empty-headedness among biologists and State Boards of Education', in *Steps to an Ecology of Mind: Collected Essays in Anthropology, Psychiatry, Evolution. and Epistemology*, University of Chicago Press, San Francisco, CA.

Bekoff, M., 2004, 'Wild justice and fairplay: cooperation, forgiveness, and morality in animals', *Biology and Philosophy*, vol. 19, pps 489-520.

Bennett, J., 2010, 'A vitalist stopover on the way to a new materialism', in D. Cole & S. Frost, *New Materialisms: Ontology, Agency, and Politics*, Duke University Press, Durham, NC.

Benson, K., 1989, 'Biology's Phoenix: Historical perspectives on the importance of the organism', *American Zoologist*, vol. 29, no. 3, pps 1067-1074.

Berger, P. & Luckmann, T., 1967, *The Social Construction of Reality: A Treatise in the Sociology of Knowledge*, Anchor Books, New York.

Bergson, H., 1998 [1911], *Creative Evolution*, A. Mitchell (trans.), Dover Publications, New York.

Berlinski, D., 1996, 'The deniable Darwin', *Commentary*, vol. 110, no. 6, pps 19-29.

Bidwell, R., 1979, *Plant Physiology*, Macmillan, New York.

Bohm, D., 1980, *Wholeness and the Implicate Order*, Routledge & Kegan Paul, London.

Bowler, P., 2009, *Evolution: The History of an Idea*, University of California Press, Berkeley, CA.

Boyd, R., 1991, 'Realism, anti-foundationalism and the enthusiasm for natural kinds', *Philosophical Studies*, vol. 61, pps 127-148.

Buckmaster, J., 1932, *Donoghue v Stevenson*, AC562:578.

Butterfield, F., 2007, 'Macroevolution and macroecology through deep time', *Palaeontology*, vol. 50, part 1, pps 41-55.

Camus, P., 2000, 'Evolution in Chile: natural drift versus natural selection, or the preservation of favoured theories in the struggle for knowledge', *Revista Chilena de Historia Natural*, vol. 73, pps 215-219.

Canfield, D., 2014, *Oxygen: A Four Million Year History*, Princeton University Press, Princeton, NJ.

Canquilhem, G., 2008, *Knowledge of Life (Forms of Living)*, S. Geroulanos & D. Ginsberg, (trans.), P. Marrati & T. Meyers, (eds.) Fordham University Press, New York.

Capra, F., 2002, 'Complexity and life', *Emergence*, vol. 4, pps 15-33.

Capra, F. & Luisi, P., 2015, *The Systems View of Life: A Unifying Vision*, Cambridge University Press, Delhi.

Carr, E., 2008, *What is History?*, Penguin Books, Melbourne, Vic.

Casti, J., 1994, *Complexification: Explaining a Paradoxical World Through the Science of Surprise*, Abacus Books, London.

Cavalier-Smith, T., 2006, part; 'Cell evolution and Earth history: stasis and revolution', *Philosophical Transactions of the Royal Society: Biology*, vol. 361, pps 969-1006.

Chambers, R., 1844, *Vestiges of the Natural History of Creation*, Churchill, London.

Cloud, P., 1972, 'A working model of the primitive Earth', *American Journal of Science*, vol. 272, pps 537-548.

Collingwood, R., 1960 [1945], *The Idea of Nature*, Oxford University Press, New York.

Collingwood, R., 1978 [1939], *An Autobiography*, Clarendon Press, Oxford.

Compton, J., 1984, 'Marjorie Grene and 'The Phenomenon of Life', *PSA Symposia*, vol. 2, pps 354-364.

Corning, P., 1983, *The Synergism Hypothesis: A Theory of Progressive Evolution*, McGraw-Hill, New York.

Corning, P., 2014, 'Evolution "on purpose": how behaviour has shaped the evolutionary process', *Biological Journal of the Linnaean Society*, vol. 112, pps 242-260.

Cosans, C., 2009, *Owen's Ape and Darwin's Bulldog: Beyond Darwinism and Creationism*, Indiana University Press, Bloomington, IN.

Darwin, C., 1859-72, *On the Origin of Species by Means of Natural Selection, or the Preservation of Favoured Races in the Struggle for Life,* John Murray, London.

Darwin, C., 1868, *The Variation of Animals and Plants under Domestication*, John Murray, London.

Darwin, C., 1872, *The Expression of the Emotions in Man and Animals*, John Murray, London.

Davies, N., 1997, *Europe: A History*, Pimlico, London.

Deleuze, G. & Guattari, F., 1993, *A Thousand Plateaus*, University of Minnesota Press, Minneapolis, MN.

Dethier, V., 1963, *The Physiology of Insect Senses*, Methuen, London.

Dobzhansky, T., 1937, *Genetics and the Origin of Species*, Columbia University Press, New York.

Escobar, J., 2012, 'Autopoiesis and Darwinism', *Syntheses*, vol. 185, pps 53-72.

Fichte, J., 2000 [1794], *Foundations of Natural Rights According to the Principles of Wissenschaftslehre*, M. Baur, (trans.), Cambridge University Press, Cambridge.

Fisher, R., 1930, *The Genetical Theory of Natural Selection*, Clarendon Press, Oxford.

Fodor, J., 1983, *The Modularity of Mind: An Essay on Faculty Psychology*, MIT Press, MA.

Forterre, P., 2005, 'The two ages of the RNA world, and transition to the DNA world: a story of viruses and cells', *Biochimie*, vol. 87, pps 793-803.

Gatlin, L., 1972, *Information and the Living System*, Columbia University Press, New York.

Gelman, R., 2009, 'Learning in core and noncore domains', in L. Tommasi *et al.*, *Cognitive Biology*, MIT Press, Cambridge, MA.

Gibson, J., 1979, *The Ecological Approach to Visual Perception*, Houghton-Mifflin, Boston, MA.

Godfrey-Smith, P., 2017, *Other Minds: The Octopus and the Evolution of Intelligent Life*, William Collins, London.

Goethe, J., 2009 [1790], *The Metamorphosis of Plants*, D. Miller (trans.), MIT Press, Cambridge, MA.

Goldenfeld, N., 2016, 'We need a theory of life', in Mazur, S., *The Origin of Life Circus*, Caswell Books, New York, pps 347-359.

Goldschmidt, R., 1933, 'Some aspects of evolution', *Science*, vol. 78, no. 2033, pps 539-547.

Goldschmidt, R., 1982 [1940], *The Material Basis of Evolution*, Yale University Press, New Haven, CT.

Goodwin, B., 1963, *Temporal Organization in Cells: A Dynamic Theory of Cellular Control Processes*, Academic Press, London.

Gordin, M., 2015, *Scientific Babel: The Language of Science from the Fall of Latin to the Rise of English*, Profile Books, [Amazon Digital Services].

Gould, S., 1994, 'The evolution of life on the Earth', *Scientific American*, pps 85-91.

Greene, M., 1982, *Geology in the Nineteenth Century: Changing Views of a Changing World*, Cornell University, Ithaca, NY.

Greene, B., 2005, *The Fabric of the Cosmos: Space, Time, and the Texture of Reality*, Vintage Books, New York.

Grene, M., 1974, *The Knower and the Known*, Universality of California Press, Berkeley, CA.

Grene, M., 1995, *A Philosophical Testament*, Open Court, Chicago, IL.

Griffin, D., 1976, *The Question of Animal Awareness*, Rockefeller University Press, New York.

Griffin, D., 2001, *Animal Minds: Beyond Cognition to Consciousness*, University of Chicago Press, Chicago.

Gross, C., 1998, 'Claude Bernard and the constancy of the internal environment', *The Neuroscientist*, vol. 4, pps 380-385.

Guthrie, S., 1993, *Faces in the Clouds: A New Theory of Religion*, Oxford University Press, New York.

Halder, G. *et al.,* (1995), 'Induction of ectopic eyes by targeted expression of eyeless gene in Drosophila', *Science*, vol. 267, pps 1788-1792.

Harris, W., 1997, '*Pax-6*: Where to be conserved is not conservative', *Proc. Natl. Acad. Sci. USA.*, vol. 94, pps 2098-2100.

Haukioja, E., 1982, 'Are individuals really subordinated to genes? A theory of living entities', *Journal of Theoretical Biology*, vol. 99, pps 357-375

Hayes, P., 2009, 'The ideology of Relativity: The case of the clock paradox', *Social Epistemology*, vol. 23, no. 1, pps 57-78.

Ho, M., 1991, 'The role of action in evolution: evolution by process and the ecological approach to perception, *Cultural Dynamics*, vol. 4, pps 336-354.

Ho, M. & Fox, S., (eds), 1988, *Evolutionary Processes and Metaphors*, John Wiley, Chichester, West Sussex.

Ho, M. & Saunders, P., 1979, 'Beyond neo-Darwinism - an epigenetic approach to evolution', *Journal of Theoretical Biology*, vol. 78, pps 573-591.

Hobhouse, L., 1926, *Development and Purpose: a Philosophy of Evolution*, Macmillan, London.

Hodos, W. & Campbell, C., 1969, '*Scala Naturae*: Why there is no theory in comparative psychology', *Psychological Review*, vol. 76, no. 4, pps 337-350.

Hoffmeyer, J., 2001, 'Seeing virtuality in nature', *Semiotica*, vol. 134, no 4, iss. 4, pps 381-398.

Hofsten, C., 2014, 'Predictive actions', *Ecological Psychology*, vol. 26, pps 79-87.

Hull, D., 1973, *Darwin and His Critics: The Reception of Darwin's Theory of Evolution by the Scientific Community*, University of Chicago Press, Chicago, IL.

Hull, D., 1974, *Philosophy of Biological Science*, Prentice-Hall, Englewood Cliffs, NJ.

Husserl. E., 1991 [1893], *On the Phenomenology of the Consciousness of Internal Time*, J. Brough, (trans.), Kluwer Academic, Dordrecht.

Huxley, T., 1852, 'Upon animal individuality: abstract of a Friday evening discourse', *Proceedings of the Royal Institution*, vol. i, pps 1851- 4.

Ingold, T., 1986, *Evolution and Social Life*, Cambridge University Press, Cambridge.

Jablonka, E. & Lamb, M., 2008, 'Soft inheritance: challenging the Modern Synthesis', *Genetics and Molecular Biology*, vol. 31, no. 2, pps 389-395.

Jablonka, E. & Lamb, M., 2014, *Evolution in Four Dimensions: Genetic, Epigenetic, Behavioral, and*

Symbolic Variation in the History Of Life, MIT Press, Cambridge, MA.

Jackson, T., 1997, 'Charles and the hopeful monster: postmodern evolutionary theory in *The French Lieutenant's Woman*', *Twentieth Century Literature*, vol, 43, no. 2, pps 221-242.

Jacob, J. & Jacob, M., 1980, 'The Anglican origins of modern science: the metaphysical foundations of the Whig Constitution', *History of Science Society*, vol. 71, no. 2, pps 251-267.

Jacob, E., Becker, I., Shapira, Y. & Levine, H., 2004. 'Bacterial linguistic communication and social intelligence', *Trends in Microbiology*, Vol. 12, no. 8, pps 336-372.

Jacob, F., 1982, *The Possible and Actual*, University of Washington Press, Seattle, WA.

Jeans, J., 1981 [1943], *Physics and Philosophy*, Dover Publications, New York.

Johnson, M., 1987, *The Body in the Mind: The Bodily Basis of Meaning, Imagination, and Reason*, University of Chicago Press, Chicago, IL.

Jonas, H., 2001 [1966], *The Phenomenon of Life: Towards a Philosophical Biology*, Northwestern University Press, Evanston, IL.

Jones, R., 1982, *Physics as Metaphor*, University of Minnesota Press, Minneapolis, MN.

Jones, C., Lawton, J, & Shachak, M., 1997, 'Organisms as ecosystem engineers', *Oikos*, vol. 69, pps 373-386.

Kahneman, D., 2011, *Thinking, Fast and Slow*, Penguin Books, London.

Kant, I., 1894 [1770], *Inaugural Dissertation: On the Form and Principles of the Sensible and Intelligible World*, W. Eckoff, (trans.), Columbia College, New York.

Kant, I., 1929 [1787], *Critique of Pure Reason*, N. Smith, (trans.), Palgrave Macmillan, London.

Kauffman, S., 1996, *At Home in the Universe: The Search for Laws of self-Organization and Complexity*, Oxford University Press, New York.

Keller, E., 1995, *Refiguring Life: Metaphors of Twentieth-Century Biology*, Columbia University Press, New York.

Kelly, G., 1963, *A Theory of Personality: The Psychology of Personal Constructs*, Norton & Company, New York.

Kingdom, J., 1993, *The Self-Made Man*, Simon & Schuster, New York.

Koonin, E. & Wolf, Y., 2016, 'Just how Lamarckian is CRISPR-Cas immunity: the continuum of evolvability mechanisms', *Biology Direct*, vol. 11, no. 9.

Kranich, E., 1999, *Thinking Beyond Darwin: The Idea of Type as a Key to Vertebrate Evolution*, Lindisfarne Books, New York.

Kropotkin, P., 1902, *Mutual Aid: A Factor of Evolution*, William Heinemann, London.

Lakoff, G., 2000, *Where Mathematics Comes From: The Embodied Mind Brings Mathematics Into Being*, Basic Books, New York.

Lakoff, G. & Johnson, M., 1980, *Metaphors We Live By*, University of Chicago Press, Chicago.

Lamarck, J., 1914 [1809], *Zoological Philosophy: An Exposition with Regard to the Natural History of Animals*, H. Elliot, (trans.), Macmillan, London.

Lane, N., 2016, *The Vital Question: Why Life Is The Way It Is*, Profile Books, London.

Lashin, S., Suslov, V. & Matsuskin, Y, 2012, 'Theories of biological evolution from the viewpoint of the modern systematic biology', *Russian Journal of Genetics*, vol. 48, no. 5, pps 481-496.

Laughlin, C. & d'Aquili, E., 1974, *Biogenetic Structuralism*, Columbia University Press, New York.

Lefebvre, V., 1985, 'The Golden Section and an algebraic model of ethical cognition', *Journal of Mathematical Psychology*, vol. 29, pps 289-310.

Lederberg, J., 1952, 'Cell genetics and hereditary symbiosis', *Physiological Reviews*, vol. 32, pps 403-430.

Levins, R. & Lewontin, R., 1985, *The Dialectical Biologist*, Harvard University Press, Cambridge, MA.

Levit, G. et al., 2008, 'Alternative evolutionary theories: A historical survey', *Journal of Bioeconomy*, vol. 10, pps 71-96.

Lewis, C., 2001, *The Abolition of Man*, Harper One, New York.

Lewontin, R., 1974, *The Genetic Basis of Evolutionary Change*, Columbia University Press, New York.

Lima-de-Faria, A., 1983, *Molecular Evolution and Organization of the Chromosome*, Elsevier Scientific, Amsterdam, NL.

Lima-de-Faria, A., 1988, *Evolution Without Selection: Form and Function by Autoevolution*, Elsevier Science, Amsterdam, NL.

Lima-de-Faria, A., 1997, 'The atomic basis of biological symmetry and periodicity' BioSystems, vol. 43, pps 115-135.

Lima-de-Faria, A., 2017, *Periodic Tables Unifying Living Organisms at the Molecular Level; The Predictive Power of the Law of Periodicity*, World Scientific Publishing, Singapore.

Lotka, A., 1945, 'The law of evolution as a maximal principle', *Human Biology*, vol. 17, no. 3, pps 167-194.

Lotka, A., 1956, *Elements of Mathematical Biology*, Dover Publications, New York.

Luisi, P., 2016, 'Origin of Life "mindstorms" needed', in Mazur, S., *The Origin of Life Circus*, Caswell Books, New York, pps 360-373.

Mameli, M. & Bateson, P., 2011, 'An evaluation of the concept of innateness', *Philosophical Transactions of The Royal Society*, vol. 366, pps 436-443.

Mannheim, K., 2015 [1936], *Ideology and Utopia: an Introduction to the Sociology of Knowledge*, L. Wirth & Shils, E. (trans.), Martino Publishing, Mansfield Centre, CT.

Marais, E., 2009 [1937], *The Soul of the White Ant*, New York University Press, New York.

Marder, M., 2012, 'Plant intentionality and the phenomenological framework of plant intelligence', *Plant Signalling and Behaviour*, vol. 7, no. 11, pps 1365-1372.

Margulis, L. & Sagan, D., 1995, *What is Life?*, Simon Schuster, New York.

Margulis, L., 1970, *Origin of the Eukaryote Cell*, Yale University Press, New Haven, CT.

Marijuán. P., et al., 2010, 'On prokaryotic intelligence: strategies for sensing the environment', *Biosystems*, no. 99, pps 94-103.

Marmet, P., 1990, 'Big Bang cosmology meets an astronomical death', *21st Century, Science and Technology*, vol. 3, no. 2, pps 52-59.

Martin, W., 2011, 'Early evolution without a tree of life', *Biology Direct*, vol. 6, no. 36.

Marx, K., 1993 [1859], *Grundisse: Contribution to the Critique of Political Economy*, Penguin Classics, Harmondsworth, Middx.

Maturana, H., 2011, 'Preface to the second edition of *Living Systems*', *Systems Research and Behavioural Science*, vol. 28, pps 583-600.

Maturana, H. & Mpodozis, J., 2000, 'The origin of species by natural drift', *Chilena de Historia Natural*, vol. 73, pps 261-310.

Maturana, H. & Varela, F., 1980, *Autopoiesis and Cognition: The Realisation of the Living*, Reidel Publishing Company, Boston, MA.

Maturana, H. & Varela, F., 1998, *The Tree of Knowledge: The Biological Roots of Human Understanding*, R. Paolucci, (trans.), Shambhala, Boston, MA.

Maturana, H. & Varela, F., 1980, *Autopoiesis and Cognition: The Realization of the Living*, Reidl, Dordrecht.

Maturana, H., 1975, 'The organization of the living: a theory of the living organization', *International Journal Man-Machine Studies*, vol, 7, pps 313-332.

Maupertuis, P., 1746, *Venus Physique: L'Origins des Hommes at des Animaux*, Jean Martin Husson, The Hague.

Mayor, A., 2011, *The First Fossil Hunters: Dinosaurs, Mammoths, and Myth in Greek and Roman Times*, Princeton University Press, Princeton, NJ.

Mayr, E., 2004, *What Makes Biology Unique: Considerations on the Autonomy of a Scientific Discipline*, Cambridge University Press, Cambridge.

Mazur, S., 2015, 'Gangen with virologist Luis Perez Villarreal', *Scoop*, July.

Mazur, S., 2016 (a), *The Origin of Life Circus: A How To Make Life Extravaganza*, Caswell Books, New York.

Mazur, S., 2016 (b), *Royal Society: The Public Evolution Summit*, Caswell Books, New York.

Mazzarello, P., 1999, 'A unifying concept: the history of cell theory', *Nature Cell Biology*, vol. 1, pps E13-215.

Meredith, J., 1928, *Kant's Critique of Teleological Judgment*, Clarendon Press, Oxford.

Megarry, J., 1979, *Malone v Police Commissioner*, 1 Ch 344.

Merker, B., 2005, 'The liabilities of mobility: a selection pressure for the transition to consciousness in animal evolution', *Science Direct, Consciousness and Cognition*, vol. 14, pps 89-114.

Merleau-Ponty, M., 1963 [1942], *The Structure of Behavior*, A. Fisher, (trans.), Duquesne University Press, Pittsburgh, PA.

Michotte, A., 1963, *The Perception of Causality*, Methuen, London.

Minot, C., 1902, 'The problem of consciousness in its biological aspects', *Science*, vol. 16, no. 392, pps 1-12.

Monod, J., 1972, *Chance and Necessity*, A. Wainhouse, (trans.), Collins, London.

Montalenti, G., 1974, 'From Aristotle to Democritus via Darwin', in F. Ayala & T. Dubzhansky, (eds), *Studies in the Philosophy of Biology*, Macmillan, London.

Moore, J., 2015, *Capitalism in the Web of Life: Ecology and the Accumulation of Capital*, Verso, London.

Morange, M., 2008, *Life Explained*, M. Cobb & M. DeBevoise, (trans.), Yale University Press, New Haven, CT.

Moser, P., 1987, *A Priori Knowledge*, Oxford University Press, Oxford.

Mumford, L., 1965, 'Technics and the nature of man', *Nature*, vol. 208, no. 5014, pps 923-928.

Nagel, T., 1974, 'What is it like to be a bat?', *Philosophical Review*, vol. 83, pps 335-450.

Nagel, T., 1986, *The View From Nowhere*, Oxford University Press, New York.

Nakagaki, T., 2001, 'Smart behavior of true slime mold in a labyrinth', *Research in Microbiology*, vol. 152, pps 767-770.

Nietzsche, F., 2009, *Writings from the Early Notebooks*, Cambridge University Press, Cambridge.

Nikolic, D., 2015, 'Practopoiesis: or how life fosters a mind', *Journal of Theoretical Biology*, vol. 373, pps 40-61.

Noble, D., 2008 (a), *The Music of Life: Biology Beyond Genes*, Oxford University Press, Oxford.

Noble, D., 2008 (b), 'Genes and causation', *Philosophical Transactions of the Royal Society*, vol. 366, pps 3001-3015.

Noble, D., 2015, 'Evolution beyond neo-Darwinism: a new conceptual framework', *Journal of Experimental Biology*, vol. 218, pps 7-13.

Noble, D., Jablonka, E., Joyner, M., Muller, G. & Omholt, S., 2014, 'Evolution evolves: physiology returns to centre stage', *Journal of Physiology*, vol. 592, pps 2237-2244.

Oyama, S., 2000, *The Ontogeny of Information: Developmental Systems and Evolution*, Duke University Press, Durham, NC.

Oyama, S., Griffiths, P. & Gray, R., (eds), 2001, *Cycles of Contingency: Developmental Systems and Evolution*, MIT Press, Cambridge, MA.

Penny, D., 2005, 'An interpretive review of the origin of life research', *Biology and Philosophy*, vol. 20, pps 633-671.

Piaget, J., 1979, *Behaviour and Evolution*, Routledge & Kegan Paul, London.

Pimentel, J., 2017, *The Rhinoceros and the Megatherium: An Essay in Natural History*, Harvard University Press, Cambridge, MA.

Pisani, F., 2007, 'Networks as a unifying pattern of life involving different processes at different levels: an interview with Fritjof Capra', *International Journal of Communication*, vol. 1, pps 5-25.

Plotkin, H., (ed.) 1988, *The Role of Behavior in Evolution*, MIT Press, Cambridge, MA.

Plotkin, H., 1996, 'The evolution of intelligence and culture', Inaugural Lecture: *International Institute of Social History*, Amsterdam.

Plutarch, M., (1957), *Moralia*: *De sollertia animalium*, (On the cleverness of animals), Vol, XII:67, Loeb Classical Library, Harvard University Press, Cambridge, MA.

Polizzotti, M., 2018, *Sympathy for the Traitor: a Translation Manifesto*, MIT Press, Cambridge, MA.

Pollan, M., 2013, 'The intelligent plant', *New Yorker*, vol. 89, no. 42, p, 92.

Popper, K., 1991, *The Open Universe: An Argument for Indeterminism*, Routledge, London.

Popper, K., 1976, 'Darwinism as a metaphysical research program', in *Unended Quest: An Intellectual Autobiography*, Fontana, London, pps 167180.

Portman, A., 1990, *A Zoologist Looks at Humankind*, J. Schaefer, (trans.), Columbia University Press, NY.

Postman, N., 1996, *The End of Education: Redefining the Value of School*, Vintage Books, New York.

Prigogine, I., 1997, *The End of Certainty: Time, Chaos, and the New Laws of Nature*, The Free Press, New York.

Pross, A., 2012, *What is Life: How Chemistry Becomes Biology*, Oxford University Press, Oxford.

Qin, J., 2015, 'Dinosaur climate probed: Ancient lake sediments in China record epic temperature swings, biotic turnover before the mass extinction', *Science*, vol. 348, no. 6240, p. 1185.

Quine, W., 1951, 'Two dogmas of empiricism', *The Philosophical Review*, pps 20-43.

Raup, D., 1972, 'Taxonomic diversity during the Phanerozoic', *Science*, vol. 177, no. 4054, pps 1065-1071.

Reed, E., 1978, 'Darwin's evolutionary philosophy: The Laws of Change', *Acta Biotheoretica*, vol. 27, no. 3, pps 201-235.

Rensch, B., 1960, *Evolution above the Species Level*, Columbia University Press, New York.

Reynolds, A., 2010, 'The redoubtable cell', *Studies in History and Philosophy of Biological Biomedical Sciences*, pps 194-201.

Riedl, R., 1983, 'The role of morphology in the theory of evolution', in M. Grene (ed.), *Dimensions of Darwinism*, Cambridge University Press, New York, pps 205-240.

Rieppel, O., 2012, 'Othenio Abel (1875-1946) and 'the phylogeny of the parts'', *Cladistics*, vol. 29, pps 328-335.

Rock, P., 1973, *Deviant Behaviour*, Hutchinson, London.

Rudwick, M., 1979, *Georges Cuvier, Fossil Bones, and Geological Catastrophes*, University of Chicago Press, Chicago, IL.

Ruiz-Mirazo, K. et al., 2004, 'A universal definition of life: autonomy and open-ended evolution,' *Origins of Life and Evolution of the Biosphere*, vol. 34, pps 323-346.

Rupke, N., 1994, *Richard Owen: Victorian Naturalist*, Yale University Press, London.

Russell, B., 1927, *An Outline of Philosophy*, Allen & Unwin, London.

Russell, B., 1997 [1927], *Religion and Science*, Oxford University Press, New York.

Russell, B., 2009 [1914], *Our Knowledge of the External World*, Routledge, Abingdon, Oxon.

Sapp, J., 1994, *Evolution by Association: A History of Symbiosis*, Oxford University Press, New York.

Schiff, M. & Lewontin, R., 1986, *Education and Class: The Irrelevance of IQ Genetic Studies*, Clarendon Press, Oxford.

Schindewolf, O., 1993 [1950], *Basic Questions in Paleontology: Geologic Time, Organic Evolution, and*

Biological Systematics, J. Schaefer, (trans.), University of Chicago Press, Chicago, IL.

Schrödinger, E., 1967 [1944], *What is Life and Other Essays*, Doubleday Anchor Books, New York.

Schlicht, E., 1998, *On Custom in the Economy*, Oxford University Press, Oxford.

Schwann, T., 1838, *Microscopical Researches into the Accordance in the Structure and Growth of Animals and Plants*, The Sydenham Society, London.

Sedgwick, A., 1859, 'From Adam Sedgwick, *Darwin Correspondence Project*, internet ref: darwinproject.ac.uk/DCP-LETT-2548.

Shapin, S., 2010, *Never Pure: Historical Studies of Science as if it Was Produced by People with Bodies, Situated in Time, Space, Culture, and Society, and Struggling for Credibility and Authority*, John Hopkins University Press, Baltimore, MD.

Shapiro, J., 2010, 'Mobile DNA and evolution in the 21st century', *Mobile DNA*, vol. 1, no. 4, pps 1-14.

Shapiro, J., 2011, *Evolution: A View From the 21st Century*, FT Press, Upper Saddle River, NJ.

Shaw, R., 2002, 'Theoretical hubris and the willingness to be radical: an open letter to James J. Gibson', *Sociological Psychology*, vol. 14, no. 4, pps 235-247.

Sheldrake, R., 2009, *A New Science of Life: The Hypothesis of Formative Causation*, Icon Books, London.

Silverton, J. & Gordon, D., 1989, 'A framework for plant behavior', *Annual Review: Ecological Systems*, vol. 20, pps 349-366.

Simon, H., 1981, *The Sciences of the Artificial*, MIT Press, Cambridge, MA.

Simpson, G., 1944, *Tempo and Mode in Evolution*, Columbia University Press, New York.

Simpson, G., 1961, *Principles of Animal Taxonomy*, Columbia University Press, New York.

Sober, E., 2008, *Evidence and Evolution: The Logic Behind the Science*, Cambridge University Press, Cambridge.

Solé, R. & Goodwin, B., 2000, *Signs of Life: How Complexity Pervades Biology*, Basic Books, New York.

Sullivan, J., 1928, *The Bases of Modern Science*, Earnest Benn, London.

Sutton, M., 2014, 'The hi-tech detection of Darwin's and Wallace's possible science fraud; big data criminology re-writes the history of contested discovery', *British Society of Criminology*, vol. 14, pps 49-64.

Thaxton, C., Bradley, W. & Olsen, R., 1984, *The Mystery of Life's Origin: Reassessing Current Theories*, Philosophical Library, New York.

Thompson, W., 1871, 'Inaugural address before the British Association at Edinburgh', *American Journal of Science*, vol. 3, no. 10, pps 269-294.

Toulmin, S., 1961, *Foresight and Understanding: An Inquiry Into the Aims of Science*, Hutchinson, London.

Toulmin, S., 1992, *Cosmopolis: The Hidden Agenda of Modernity*, University of Chicago Press, Chicago, IL.

Trewavas, A., 2014, *Plant Behaviour and Intelligence*, Oxford University Press, Oxford.

Trewavas, A. & Baluska, F., 2011, 'The ubiquity of consciousness: the ubiquity of consciousness, cognition and intelligence in life', *European Molecular Biology Association*, vol. 12, no. 12, pps 1221-25.

Turner, B., 2009, 'Epigenetic responses to environmental change and their evolutionary implications', *Philosophical Transactions of The Royal Society*, vol. 364, pps 3403-3418.

Turner, J., 2017, Review of D. Haraway, 'Life with Ms Cayenne Pepper', *London Review of Books*, vol. 39, no. 11, pps 23-27.

Turvey, M., Shaw, R., Reed, E. & Mace, W., 1981, 'Ecological laws of perceiving and acting: In reply to Fodor and Pylyshyn (1981)', *Cognition*, vol. 9, pps 237-304.

Uexküll, von J., 1926, *Theoretical Biology*, D. Mackinnon, (trans.), Harcourt Brace & Company, London.

Uexküll, von J., 2010 [1934], *A Foray into the Worlds of Animals and Humans: With a Theory of Meaning*, J. O'Neil, (trans.), University of Minnesota Press, Minneapolis.

Uexküll, von J., 1982, 'Introduction: The Theory of Meaning', *Semiotica*, vol. 42, no. 1, pps 25-82.

Vaihinger, H., 2009 [1925], *The Philosophy of 'As If': A System of the Theoretical, Practical and Religious Fictions of Mankind*, C. Ogden, (trans.), Martino Publishing, Mansfield Center, CT.

Vane-Wright, R., 2014, 'What is life? And what might be said of the role of behaviour in its evolution?', *Linnean Society Biological Journal*, no. 112, pps 219-241.

Varela, F. & Shear, J., 1999, *The View From Within: First-Person Approaches to the Study of Consciousness*, Imprint Academic, Bowling Green, OH.

Varela, F., Thompson, E. & Rosch, E., 1993, *The Embodied Mind: Cognitive Science and Human Experience*, The MIT Press, Cambridge. MA.

Villarreal, L., 2005, *Viruses and the Evolution of Life*, ASM Press, Washington, DC.

Voltaire, 1765, 'Fragments sur l'Histoire', *Euvres*, vol. 27, no, 1, pps 158-159.

Waddington, C., 1942, 'Canalisation of development and the inheritance of acquired characters', *Nature*, no. 3811, pps 563-565.

Wake, D., Roth, G. & Wake, M., 1983, 'On the problem of stasis in organismal evolution', *Journal of Theoretical Biology*, vol. 101, pps 211-224.

Wallin, I., 1927, *Symbionticism and the Origin of Species*, Williams & Wilkins Company, Baltimore, MD.

Ward, P. & Kirschvink, J., 2015, *A New History of Life: The Radical New Discoveries About the Origins and Evolution of Life on Earth*, Bloomsbury, London.

Webster, G. & Goodwin, B., 1997, *Form and Transformation: Generative and Relational Principles in Biology*, Cambridge University Press, Cambridge.

Weismann, A., 1893, *The Germ-Plasm: A Theory of Heredity*, Charles Scribner, London.

Weiss, K., 2012, 'To understand the Baboon', *Evolutionary Anthropology*, vol. 21, pps 131-5.

Weiss, K., Buchanan, A. & Lambert, B., 2011, 'The Red Queen and her King: cooperation at all levels of life', *Yearbook of Physical Anthropology*, vol. 54, pps 3-18.

West-Eberhard, M., 2003, *Developmental Plasticity and Evolution*, Oxford University Press, New York.

Whitehead, A., 1968, *Nature and Life*, Greenwood Press, New York.

Woese, C. 2004, 'A new biology for a new century', *Microbiol. Mol. Biol. Rev.* vol. 68, no. 2, pps 173-186.

Wolfe, C., 2010, 'Do organisms have an ontological status?', *History and Philosophy of the Life Sciences*, vol. 32, no. 2/3, pps 195-231.

Wolpert, L., 1995, 'Evolution of the cell theory', *Philosophical Transactions: Biological Sciences*, vol. 349, no. 1329, pps 227-233.

Wright, J. & Jones, C., 2006, 'The concept of organisms as ecosystem engineers ten years on: Progress, limitations, and challenges', *Bioscience*, vol. 56, no. 3, pps 203-209.

Zimmer, C., 2018, *She Has Her Mother's Laugh: the Powers, Perversions and Potential of Heredity*, Picador, London.

www.ingramcontent.com/pod-product-compliance
Lightning Source LLC
Chambersburg PA
CBHW021406210526
45463CB00001B/246